Pacific Crest School

THE RIVER THAT MADE SEATTLE

THE RIVER THAT MADE SEATTLE

A HUMAN AND NATURAL HISTORY
OF THE DUWAMISH

BJ Cummings

UNIVERSITY OF WASHINGTON PRESS

Seattle

www.tulalipcares.org

The River That Made Seattle was supported by a generous grant from the Tulalip Tribes Charitable Fund, which provides the opportunity for a sustainable and healthy community for all.

Design by Katrina Noble
Composed in Adobe Caslon Pro, typeface designed by Carol Twombly

24   23   22   21   5   4   3   2

Printed and bound in the United States of America

UNIVERSITY OF WASHINGTON PRESS
uwapress.uw.edu

LIBRARY OF CONGRESS CATALOGING-IN-PUBLICATION DATA ON FILE

ISBN 978-0-295-74743-9 (hardcover)
ISBN 978-0-295-74744-6 (ebook)

Cover design: Katrina Noble
Cover illustrations: (*photograph*) Ann Rasmussen, great-granddaughter of Quio-litza (aka Ann Kanum), holding the family's heirloom canoe paddle. Courtesy of Jeff Corwin. (*map*) Lionel Pincus and Princess Firyal Map Division, The New York Public Library. "Vicinity map, Duwamish River & its tributaries, Washington" New York Public Library Digital Collections. Accessed March 24, 2020. http://digitalcollections.nypl.org /items/640f77e0-2c61-0136-6f21-0bfd557ad2c4.
Frontispiece: Duwamish Waterway. Courtesy of Tom Reese.

The paper used in this publication is acid free and meets the minimum requirements of American National Standard for Information Sciences—Permanence of Paper for Printed Library Materials, ANSI Z39.48–1984.∞

To the people of the Duwamish River—past, present, and future

We are still here.

CECILE HANSEN, CHAIRWOMAN, DUWAMISH TRIBE

It's a river, not a waterway.

TIM O'BRIEN, RESIDENT/ACTIVIST, GEORGETOWN, SEATTLE

CONTENTS

PREFACE

The Duwamish is our river. Whether you live in my hometown of Seattle or in Chicago, Newark, Mobile, Des Moines, or Los Angeles, you have a river with a story, and that story resembles the one in this book. The details vary, but the broad strokes are strikingly consistent.

Across America, our nation has corralled and manipulated waterways for the prosperity of our cities, farms, and industries. We have commandeered their channels as repositories for our waste, reservoirs for our irrigation systems, and raw energy for our power plants. We have built our urban centers, expanded our agriculture, and fueled our industrial growth by exploiting our rivers' resources. In doing so, we have enriched ourselves. In every instance, we have also caused suffering and deprivation.

It is well understood that we have sacrificed whole populations of aquatic birds, fish, mammals, and plants by altering our rivers. But we have also impoverished entire populations of Indigenous peoples of the Southwest who rely on the cultural and subsistence resources of the Rio Grande River; low-income families who fish to survive on the Anacostia River in Washington, DC; and industrial fenceline communities in Pittsburgh and Cincinnati who live next to polluting factories on the Ohio River. The choices we have made about how we use our rivers reflect the values of the governing bodies of our cities and townships at the moments when those choices were made. Few of us know much about that history.

Without knowing our history, we cannot understand how it has influenced the river communities we know today. Nor can we understand history's influence on the choices we continue to make. Without knowing our history, we are doomed, as the saying goes, to repeat it—or, worse, to exacerbate its harms.

Increasing conflicts over water resources in the United States in recent years are refocusing attention on the stories of our rivers and their people. Our nation's history, including the colonial-era appropriation of North America's waterways—often by force—is showing its stress lines. In the arid West, alliances of Native tribes and environmentalists are being pitched against the interests of farmers as some states consider dismantling century-old dams. On the Sioux Nation's Standing Rock Reservation, tribal leaders led a nine-month standoff with Energy Transfer Partners and the State of North Dakota in 2016 in an attempt to protect the nearby Missouri River from pipeline spills. And in British Columbia, Wet'suwet'en hereditary chiefs are leading a fight to prevent a pipeline from crossing their autonomous lands, in part to protect the Morice River.

What do these movements and eruptions of civil unrest have in common with the struggle for control over water resources on the Duwamish and hundreds of other rivers nationwide? In many of these watersheds, communities are more quietly, though no less desperately, fighting similar threats to their health, culture, and livelihood.

By some accounts, Seattle's Duwamish River has emerged as an example of best practices in restoring the environment and promoting social justice while cleaning up past pollution. In other circles, Seattle's river cleanup is seen as a farce, driven by a powerful confluence of government and industry forces aligned against tribal and immigrant communities. The truth, and the lessons we can learn from the Duwamish case, reflects a bit of both. Grounded in colonial appropriation and fueled by contemporary racism and anti-immigrant sentiment, the treatment of the Duwamish River and its communities today leaves much to be desired, yet it surpasses the low standards set for river restoration and cleanup across much of the country.

The Duwamish story is one case study in the national effort to express our values in the way we treat our rivers and their people. The Standing

Rock battle cry—"Mni wiconi," or "Water is life"—captures the threat many communities perceive in sacrificing our rivers for national progress and financial gain. As we begin to restore riverbank habitat and to scrub decades of chemical waste from our river bottoms, we have the opportunity to act in accordance with our values. If we do enough to create a result pleasing to the eye, but insufficient to protect the health of our river-dependent communities, that decision will speak volumes about the classism and racism that underpin it. And if we demand a pristine restoration of a romanticized past, we may disenfranchise exactly those people from whom our rivers were appropriated in the first place.

Collaboration, respect, and justice are core values that we may or may not choose to guide our efforts at environmental restitution, but they are most certainly the only path forward if we want to ensure that our actions make the Duwamish, Hudson, Cuyahoga, Mississippi, and Potomac into rivers that serve all the people who live, work, fish, play, and pray in and along their waters.

ACKNOWLEDGMENTS

I extend my deepest gratitude to the people who shared their personal and family stories with me as I explored the experiences of the Duwamish River's Native, immigrant, and industrial communities. I am eternally grateful to James Rasmussen, Virginia Nelson, and the extended Kanum-Tuttle family for sharing their family albums, records, heirlooms, and histories as I followed their Duwamish family through seven generations of interactions—often painful—with the watershed's settler families and colonial governments. I also owe many thanks to Cecile Hansen and Ken Workman, who contributed valuable insights into their own experiences and those of their Duwamish ancestors. Other people who made invaluable contributions to this book include Louise Jones-Brown of the Maple family, Suzanne Hittman of the Desimone family, the extended Hamm and Medina families, Marianne Clark, Carmen Martinez, David Yamaguchi, Liana Beal, Paulina Lopez, Peter Quenguyen, and Sophorn Sim. Their stories of immigration and honest reflections about the imprint their families made on the river—and vice versa—were critical to painting the watershed's portrait.

I am also enormously thankful to the many subject-matter experts who gave of their time and knowledge as I traced changes to the historic and contemporary rivers of the Duwamish watershed: Shawn Blocker, George Blomberg, Sarah Campbell, Dan Cargill, Lee Dorrigan, Jonathan Hall, William Kombol, Anna Marti, James Meador, David Munsell, Alberto

Rodriguez, Cliff Villa, Greg Wingard, and many, many others. The librarians and archivists of this region are a treasure without whom much of our local history would be lost. I offer special thanks to the staff of the White River Museum, the Renton History Museum, the Tukwila Historical Society, the Friends of Georgetown History, the Museum of History and Industry, and the public archives of Seattle, King County, and Puget Sound, as well as the Seattle Public Library and the University of Washington Library system. I am blessed to live in a region that values historical books, archives, and public records—and also recognizes their limitations. I could not have cracked open the treasure chests of these records without the brilliant and talented research assistance of Jennifer Smith.

Finally, I am thankful for the shoulders I stand on, albeit precariously. Without doors opened by the research and writing of Mike Sato, Matthew Klingle, Coll Thrush, David Williams, Thomas Speer, and David Buerge, this book would not have been possible. The photographers and artists whose work is featured in this book generously shared their keen eye and talents to illustrate these pages: Paul Joseph Brown, Jeff Corwin, Gene Gentry McMahon, Tom Reese, Steve Shay, and Julie Whitehorn. 4Culture provided invaluable support as I developed the manuscript and began to tell the story of Seattle's river through lectures, blogs, and watershed tours. And of course, I owe the greatest debt to my personal editor, Gillian Culff, and my editors at the University of Washington Press who ushered this project from inception through production: Catherine Cocks and Andrew Berzanskis.

Above all, thank you to my family—Tom, Maya, and Marilynn—for their enduring patience, support, and encouragement. I am forever grateful.

THE RIVER THAT MADE SEATTLE

Introduction

A BULLDOZER WAS CHURNING UP THE GROUND WHERE CECILE Maxwell's ancestral village had once stood. Maxwell, the great-great grand-niece of "Chief Seattle" and the new chairwoman of the Duwamish Tribe, had been visiting the site upriver of the West Seattle Bridge often. In the months leading up to this day in early July 1976, she had frequently talked with the archaeology students sifting through carefully excavated bores of black dirt specked with shell and bone fragments. She was eager to learn about their finds and how they might affect her tribe's claims to land and fishing rights in the area.

No students were working: their project had been wrapped up, the grids and tools they'd been using in their research removed. The site had been quiet for several weeks, but today a bulldozer was working the area where the students had meticulously documented fragments of bone, stone tools, and myriad seashell deposits. Alarmed, Maxwell hurried back to her office and dialed the number for the Army Corps of Engineers.

The previous fall, Maxwell had received a letter notifying her that the Army Corps had found evidence of a tribal settlement on land owned by the Port of Seattle. The Corps's district archaeologist, David Munsell, had been reviewing an application from the port for a permit to fill a river bend that was left behind when the Duwamish Waterway was constructed more than half a century earlier. The remnant branched sharply off the deep, straight shipping channel and looped west before arcing back to rejoin the constructed waterway half a mile downriver—or a mile along the meander formed by the natural river bend.

Since 1899, Section 10 of the federal Rivers and Harbors Act had required a permit for any dredging, filling, or excavation in navigable waters of the United States. The entire Duwamish Waterway had been constructed in the early 1900s under Section 10 permits. In 1975, the port moved to fill this last remaining river bend with an eye to adding more land along the Duwamish Waterway—land that could be used to build a new marine terminal.

The port's permit application was routine, but a new Washington State law, passed in 1975, declared a state interest in protecting archaeological resources for their historical and scientific value. The Corps of Engineers had never examined any Duwamish River sites for their archaeological value before, but the port's Terminal 107 property, which sat along the remnant stretch of river, was right across the channel from Munsell's office: he could see it from the windows of the Army Corps building. So, with the application in hand, he drove across the West Seattle Bridge to take a look.

Munsell passed the Seaboard Lumber Mill before turning into a small lot at Terminal 107. He parked his car and walked to a cluster of houses in the process of being demolished. To prepare for developing the property, the port was evicting the occupants of an entire neighborhood of modest homes that stretched along the river bend. Some of the houses had already been removed, their shallow foundation pits exposed. Peering down at the exposed layers of earth in one of the pits, Munsell immediately knew that he would not be approving the port's application. A swath of exposed shell and bone fragments more than a foot deep cut across the face of the dirt— a classic midden, or disposal ground, of a type commonly associated with prehistoric village sites in coastal areas. The Port of Seattle's Terminal 107 had archaeological resources in abundance, lying bare for all to see.[1]

By the time the Port of Seattle signed a contract with the University of Washington to investigate the ancient site—a requirement of state law and a condition of the Army Corps's further consideration of their permit— Cecile Maxwell and the Duwamish Tribe had been alerted to the finding. As original occupants of the Duwamish River Valley and its network of villages, they were included in the chain of notifications issued by the Army Corps and the Port of Seattle. Maxwell had become the tribe's chairwoman a few years earlier, spurred by indignation over her brother's arrest for fishing in the Duwamish River at a time when Native fishing rights were routinely ignored.

The day she came upon the bulldozer desecrating the remains of her ancestors' village, the university archaeologists had just submitted their findings to the Port of Seattle. The shell midden that David Munsell had spotted on his visit had been confirmed as part of a site with great archaeological significance, and the university team called for further study before any development of the area. The report recommended that the site "be actively protected from any further disturbance." Nevertheless, a few weeks later the port ordered the demolition of several condemned houses on the property, right in the middle of the study area. The incident destroyed much of the documented village site before Maxwell's frantic call to the Army Corps could stop it.[2]

According to Tom Lorenz, the university's lead archaeologist at the site, the remains had been destroyed. "I'm sort of overcome by how much is gone," he told the *Seattle Times*. "This area has been so disturbed that there is very, very little left that's of use." The port insisted the demolition was accidental, but an irate letter from Washington State's historic preservation officer, Art Skolnick, accused the port of having "willfully altered this significant archaeological site" and said the bulldozers had "irrevocably destroyed a prime source of scientific data." The digging and compaction destroyed 80 to 90 percent of the known archaeological remains.[3]

Had that been all there was to the story, Terminal 107 might be a bustling marine terminal today. Heavily disturbed archaeological resources, no matter how significant and regardless of the cause of their disturbance, are not generally considered worth saving. But fortunately for the Duwamish Tribe and others with an interest in Seattle's cultural heritage, the Army Corps demanded a new study to examine the rest of the port property, to determine whether any more artifacts remained outside the disturbed area. The new study, released in March 1977, reported that while the demolition had destroyed the half acre that made up the original research area, an additional two and a half acres of archaeologically important resources were found, dating back 1,400 years. They included shell, bone, and stone tool fragments, along with the remains of an "aboriginal house structure"—possibly one of the Duwamish Tribe's ancestral longhouses or perhaps a fish-drying shelter used during the winter salmon runs. The new study recommended that the site—bureaucratically referred to as Duwamish No. 1 Site—be nominated for the National Register of Historic Places and that immediate steps be taken to ensure its protection.[4]

Most significantly, the research team concluded that "any future development activities . . . would probably result in significant adverse impacts to its cultural resources." The researchers recommended additional excavation in order to recover valuable scientific and cultural information still buried beneath the surface.[5]

In a 2017 interview, Munsell recalled that the Port of Seattle was not pleased with the outcome: "They were unaccustomed to complying with antiquities legislation." Munsell and Art Skolnick became so concerned about the port's actions and the political pressure it might exert to get its permits approved that they took the highly unusual step of alerting the news media about their findings.[6]

Public support for preserving the site erupted. While Munsell was a federal employee and somewhat shielded from the local political blowback that followed, Seattle's mayor, Wes Uhlman, reportedly complained about Skolnick's actions to the state governor, Dixie Lee Ray. Skolnick, a state employee, was soon out of a job.[7]

Despite the drama playing out behind the scenes, the Army Corps pushed forward. By the time Sarah Campbell of the University of Washington took over investigating the history of the site in the 1980s, the port had gone on record as supporting the effort. Campbell describes the port staff as being very cooperative, despite their ongoing interest in developing the site for commercial use. But as more artifacts were found, each consecutive investigation added to the public's understanding of the significance of the site and led to new recommendations for further excavation and study. By the end of the 1980s, the port had cut its losses and laid a layer of protective soil over the historic village to preserve its contents, opening up the river-bend property to public recreational access. The Duwamish Tribe's village would not be erased to make way for a new shipping terminal.[8]

Today, public art installations and interpretive history signs dot a pedestrian walking trail along the riverbank where the Duwamish Tribe's

Aerial view of the Duwamish Waterway and Kellogg Island, showing the original river bend and straightened canal, with downtown Seattle in the background. Courtesy of Tom Reese.

longhouse stood at the center of the village called Yuliqwad—Lushoot-seed for "basketry hat," a traditional cedar headpiece. Adjoining the site is a city park, Herring's House, named after another tribal village once located about a mile away, on the shore of Elliott Bay. Today, everything up- and downriver of the historic village has been altered by the construction of the Duwamish Waterway, but the river bend itself—the last remnant of the original river within the Seattle city limits—remains the same.[9]

∞

I first saw the river bend at Yuliqwad from a kayak in the spring of 1994, my center of gravity below the waterline as I cut through the narrow channel. A concave wall of ash-fine sand rose from the water on my left, binding together a layer of ancient clamshells still visible in the eroded bank. Belted kingfishers trilled as they skipped from tree to tree ahead of my boat, and a great blue heron skimmed low, flushed from the reeds on my right. Ahead, a mudflat extended around the bend, alive with speckled shorebirds scurrying along the water's edge. I felt I had passed into another time, one before barges and smokestacks and sewer grates.

Ten minutes earlier, I had been leading a team of kayakers along Seattle's concrete-lined Duwamish Waterway in a training session for volunteers with the Puget Soundkeeper Alliance's Kayak Patrol—a citizen navy of sorts, ferreting out illegal dumping and industrial discharges fouling Seattle's waters. Airplanes and cranes cast shadows over the water, and the din of engines and machines drowned out our shouted attempts at conversation. Paddling with cameras and notepads tethered to our boats, we were searching for, and finding, unauthorized sources of pollution.

Before that day, I had run my training sessions on Seattle's Lake Union and the Ship Canal that leads from the lake to Puget Sound, passing through the Ballard Locks. I had been on the Duwamish River only once before, with Lee Moyer, a former Boeing engineer who made a postretirement career of designing and building kayaks. We had stuck to the wide shipping channel that day, getting familiar with the many factories and pipes that lined the waterway. The channel had been built straight, dredged deep, and lined with industrial berths. There was nothing natural about it.

On this training excursion, I decided to hang a hard left after leading the volunteers past the last of a line of factories. "This one is Holnam Cement," I shouted over the churning gears above. A conveyor belt was offloading heaps of sand and gravel from a sixty-foot barge tied to its berth. "If you see this stuff going in the river, take a photo and grab a water sample so we can test it for pH and arsenic. The raw material they use here can be toxic to fish." Then, with twenty minutes to kill before our session ended, I led the group behind the remnant of a small island just north of the factory.

The serene meander we found ourselves paddling along, and its sudden proliferation of wildlife, was a revelation. It was here that Cecile Maxwell had encountered the bulldozers in 1976. Here early American settlers had built a bustling neighborhood of gridded streets and fishing docks in the 1890s. And here the Duwamish Tribe had lived in a cluster of longhouses

Cecile Hansen, chairwoman of the Duwamish Tribe, wearing a traditional basketry hat in 2010. The Duwamish River is in the background. Courtesy of Steve Shay.

and hosted potlatches for over a thousand years, leaving behind a dense detritus of clamshells. Maxwell's desperate call saved the last remaining bend on the Duwamish River where we now floated at eye level with the shorebirds.

In the years following, I became the director and head "Soundkeeper" at the Puget Soundkeeper Alliance, filed and resolved several lawsuits against Duwamish River industries that we caught violating the Clean Water Act, and founded the Duwamish River Cleanup Coalition, an organization focused on securing a thorough cleanup of the toxic pollution that had accumulated in the river bottom over the past hundred years. In time, I came to be considered an expert on the Duwamish River—Seattle's hometown and only river, but one which few people in the city knew much about, if they even knew it existed. Unless you had reason to be on the river itself—and few outside the adjacent industries and shipping did—the river remained almost entirely hidden behind an industrial wall.

Researching this book confirmed what I had always suspected: that despite twenty-five years advocating for the Duwamish and working closely with those who depend on and love it best, I hardly knew this river at all. The history of the Duwamish River is deep, rich, and long—and as complex as the stories of the many generations of Native and immigrant families who have now settled here from every continent and multiple archipelagoes around the world. In the past two centuries, we have dramatically changed the course of this river, but not so dramatically as the transformations we wrought have in turn changed us.

∞

The story of the Duwamish River and the experiences of its people—Native, immigrant, and industrialist—is largely missing from the popular history of Seattle. The river's original watershed extended from Mount Rainier's Emmons Glacier to the north King County suburb of Woodinville and included the White, Green, Black, and Cedar Rivers, Lakes Washington and Sammamish, and a spiderweb of interconnected creeks and lakes, from north Seattle's Green Lake to Roaring Rock Creek in southwest King County and all the ponds, wetlands, and tributaries in between. The entire watershed drained through the Duwamish River to the Puget Sound embayment we call Elliott Bay, on downtown Seattle's waterfront. It was the land of the Dkhw'Duw'Absh or Doo-Ahbsh ("people of the inside") and the closely related Hah-chu-Ahbsh ("lake people"), today collectively known as the Duwamish Tribe.

The changes to the watershed did not begin with the arrival of the Denny Party, commonly believed to be the city's first settlers, but with the very first white immigrants to the area now known as Seattle: Jacob and Samuel Maple, Henry van Asselt, and the Collins family. Nearly two weeks before the 1851 visit by the Denny Party's scouts and two months before the rest of the Denny Party landed at Alki Point, these settlers completed their journey from the goldfields of California, through present-day

Olympia, to the shores of the Duwamish River. After a foray into California gold mining, Luther Collins abandoned his farm on the Nisqually River, near the British-owned Hudson Bay Company's trading post. He joined a trio of other travelers to scout out a new destination a full day's paddle north of the company post. Collins, who had visited this "unsettled" river before, stoked his new companions' ambitions with descriptions of the fertile Duwamish Valley as an ideal homeland with friendly natives. The settlers of the 1850s named their new city Seattle, for the tribal leader who welcomed and supported them when they arrived.[10]

Since this pioneering party first settled on the Duwamish River, alliances and conflicts between and among Native peoples, immigrant residents, and local and global industrialists have transformed the watershed's natural resources, its economy, and all of its communities. The City of Seattle grew from the rich resources of the river's tide flats, from the monumental feats of its early industrial barons, and from the persistence of generations of Native and immigrant residents. But this growth came at a high cost.

Only seventy years after the first colonists settled on the Duwamish River, its watershed had been reduced to less than one-quarter of its original size of more than two thousand square miles, and only the waters of the Green River still flowed freely to the Duwamish. The White, Black, and Cedar Rivers had been diverted to bypass the Duwamish or had dried up entirely. The waters of the freshwater lakes that these rivers fed and drained were forced through newly engineered routes. The Native people who lived by the changed rivers had been similarly "diverted" to reservations, relegated to shantytowns, integrated into settler society through marriage, or eliminated through disease and warfare.

As an engineering feat, the transformation was remarkable. The dramatic alterations to the Duwamish watershed, and to the river itself, allowed for the birth of a thriving industrial city. Business boomed. Immigrants flocked to the growing metropolis from all corners of the world. From the banks of the Duwamish, a city was born.

And there the story we tell ourselves about the making of Seattle typically ends. But we've neglected to pursue important questions about who, and what, we became. What happened to the people who were in the path of this progress? Where are the surviving families of the Duwamish Tribe today? Who are the descendants of those first immigrants to settle on the river? Who has taken their place? And what became of the riverfront businesses founded by Seattle's industrial pioneers?

Today the Duwamish River is polluted, its neighborhoods in poor health, and its industrial base struggling. At the start of the twentieth century, the city's boosters filled the mudflats at the mouth of the river to create one of the world's largest artificial islands. In 1913, dredgers began to straighten the river's bends and deepen its draft for easy access by ships. The land bordering this new channel was leveled and filled as a site for factories in an effort to create a modern industrial city. By the time the valley filled with the noisy bustle of commerce and industry, less than 2 percent of the river's original habitat remained, pushing local salmon runs and wildlife close to extinction. For the rest of the century, the river was used as a waste repository, until the federal government declared the river a Superfund site—one of the nation's most hazardous waste sites— and ordered a cleanup. News of this directive was published in the *Seattle Post-Intelligencer* on September 14, 2001—150 years to the day after the first settlers arrived.

The studies that followed the cleanup order revealed a legacy of water, land, and air pollution with tragic health consequences for local residents and fishermen. Land and business values stagnated as more contamination was discovered, and the full cost of cleanup—and liability—skyrocketed. Today, we stand at the precipice of the decisions that may, or may not, restore the health of the Duwamish River and its diverse communities.

This book traces the environmental and social history of Seattle's home-town river. It tells the stories of the lost rivers and lakes of the Duwamish watershed, the people who helped destroy them, and those who suffered from their loss. It traces the families of the region's first immigrants and the history of the more recent arrivals—immigrants from Italy, Japan, Eastern Europe, Mexico, Central and South America, Southeast Asia, East Africa, and other places who have become a lasting part of a city long regarded as overwhelmingly white. It also follows the Native people who fled, fought or married the new immigrants, and the industrialists who fueled the growth of the city. And it shows that friends, neighbors, and colleagues in twenty-first-century Seattle are often the descendants of all three.

The book also describes those who are still here—notably, the families of Se'alth and his allied chiefs, as well as the Maples, Collinses, Van Asselts, and their early neighbors. They continue to shape our city, and their descendants are an integral part of its still-unfolding history.

A century and a half after Chief Se'alth escorted the first pioneer families upriver, many of these people have come together in a bid to reclaim Seattle's river. Whether they will succeed depends on many factors, one of which is whether Seattle can agree on an answer to the question that has been debated for decades: Whose river is it? Plans for cleanup and revitalization are under way, led by this new generation of pioneers who reimagine the Duwamish as "a river for all."

1

In the Beginning

The "Duwamps" and Its First People

BEFORE 1850

STANDING ON A MORAINE OF SHALE AND BLACK SAND ABOVE THE river valley, I have a direct view into the ice caves at the base of the Emmons Glacier. My hiking companion and I have trekked here to see the birthplace of the White River. The sheer mass and sense of power locked in the expanse of ice cascading from the summit of Mount Rainier is overwhelming. The river flows from the caves, tucked under a grey-brown layer of dirt and rock that was dumped by a massive landslide in 1963. Above the slide zone, the glacier's blue-tinged ridges are broken by rock outcroppings and jagged peaks. Below the slide, the ice is slowly advancing—one of the last remaining glaciers that is growing, rather than retreating—thanks to the insulating power of the blanket of rock. The White River is true to its name, colored by its load of cloudy glacial silt. It splashes over and around the rocks and fallen logs in its path as it tumbles into the valley below.[1]

Of the twenty-five glaciers covering thirty-five square miles of Mount Rainier, the Emmons Glacier is the largest. It once extended far down into the foothills and valleys, merging with other ice sheets and feeding an immense freshwater lake in the depression that is now Puget Sound. As

View of Emmons Glacier on Mount Rainier. The White River emerges from below the rubble of the 1963 landslide at the base of the glacier, lower left. Courtesy of Julie Whitehorn.

the climate warmed and the glaciers retreated, the runoff from the Emmons Glacier became the source of the White and Duwamish Rivers. This transformation, which took millennia, included a period of glacial advance and retreat that is remembered in the legends of the region's first peoples.[2]

ORIGINS

Unbroken ice sheets up to a mile thick covered the area we call Puget Sound. Between eighteen thousand and sixteen thousand years ago, the glaciers began a spasmodic retreat from the region. An irregular warming trend pushed the boundary of the glacial front north, sculpting a new landscape as the tendrils and lobes of the ice cut deep into the earth. Frozen ground began to warm and breathe again for the first time in millennia, but much of the area remained underwater. Until the glacier retreated north of the Strait of Juan de Fuca, its meltwater was trapped, forming huge lakes. Eventually, the backed-up freshwater was able to escape through the strait. In its place, the Pacific Ocean flowed in, and an enormous inland sea filled the area from today's Admiralty Inlet to Commencement Bay. Sediment deposition and mud flows from volcanic lahars slowly filled the shallows and created new landforms.[3]

As the land warmed, sprouts of green emerged, and wildlife began to venture into the new terrain. The inevitable press of human hunters followed, exploring as far as the shifting ice would allow. Retreating glaciers and new deposition created river valleys and wide plains that were settled by the recently uncovered land's first peoples.

These new arrivals entered the Pacific Northwest from across the Bering Strait more than twelve thousand years ago, venturing south until they were in land vacated by the retreating "Puget Lobe" of ice. The first recorded settlements in the region thrived on the Sammamish River at Bear Creek, near present-day Woodinville.[4]

Native people's migration into this emergent landscape is captured in their founding stories. The "Epic of the Winds" chronicles the retreat of the ice fields from the Duwamish River Valley: "The land was locked in ice. Every day was bitter cold. North Wind sent his freezing blast all over the country. He built an ice dam across the Duwamish River that stopped the salmon from running upstream, so the people were always hungry and cold. Month after month, nothing ever changed."[5]

The legend describes how the ice was driven from the valley. North Wind's jealous love for a woman called Mountain Beaver drove him to kill her husband, Chinook Wind. In her grief, Mountain Beaver hid and gave birth to her husband's son, Storm Wind. Years later, the boy avenged his father's death by driving North Wind away. "Storm Wind blew the rain all over the valley. Now, everything was melting. All the land was becoming unfrozen. North Wind ran away. He went further north. . . . If [he] had not been chased away we should all be cold and hungry all the time. As it is, we have a little ice and snow, but not for long, only until Storm Wind comes again."[6]

This and other stories record changes in the geological record that predate written accounts. Oral histories of tribes throughout the region tell of a battle between Thunder Bird and a supernatural rival, which caused the shaking of the ground and a "rolling up of the great waters"—a reference to an earthquake and tsunami on the Northwest coast in 1700 that were felt as far away as Japan. With no literate societies in the Northwest at the time, the orally transmitted legends and the land itself have preserved the record of what occurred.[7]

Anthropologists have unearthed evidence of human occupation in the northern Duwamish watershed twelve thousand years ago, and most of the river valleys were settled between seven thousand and eight thousand years ago. The first accounts from the early years of exploration contradict many of the stories Americans have told themselves and their descendants about their own arrival and early years in Puget Sound. These stories are not widely known, but they are key to a more accurate and full telling of our history.[8]

FIRST ENCOUNTERS

In the watery expanse of Puget Sound in the 1780s, a young Suqw-Ahbsh (Suquamish) leader paddled across the wide expanse of water—the Whulge—that separated his island from the mainland. He continued up a narrowing river delta to the village of Stuk to marry his bride. By some accounts, she had been a slave, while others report that the woman was a high-born, or "royal" member of the Doo-Ahbsh (Duwamish) tribe, daughter and sister to leaders of the village. The newly married couple, Schwabe and Sholitza, gave birth to a son, Se'alth, who was to become the best-known of the region's Native leaders, Chief Seattle. Seven generations later, Se'alth's descendants are the keepers of his family's story and the story of his birth on nearby Blake Island in 1786.[9]

Another high-born Duwamish woman, Tupt-Aleut, from the village of Sbabadid on the Black River, married Kruss Kanum, the son of a prominent Skagit chief from Whidbey Island. These intertribal family alliances opened trade relationships and ensured access to resources on both sides of Puget Sound, including roots and berries from the lowland, deer and elk from the mountains, and fish, shellfish, birds, and small mammals near the sound.[10]

Kruss Kanum succeeded his father as chief. He and Tupt-Aleut greeted the first settlers to Whidbey Island at the same time Se'alth was welcoming settlers to present-day Seattle. Following their long-standing tradition, many Native leaders and their villages met, hosted, and cultivated relationships with the first traders and settlers to reach Puget Sound. While intentional, these relationships were often strained, and the alliances forged were complicated and uneasy. But following local traditions, these bonds were sometimes strengthened through marriage. The families of Se'alth and Kanum pursued these alliances, and both chiefs led their people through the changes that followed.[11]

A decade before Se'alth's birth, on the other side of the continent, American colonists had declared their independence from Britain and set their sights on nation building and westward expansion. The first written descriptions of the Pacific Northwest's interior waters reached them in 1792. For decades, European traders had been exchanging goods with Northwest coastal tribes. Nootka Sound, halfway up Vancouver Island's rugged west coast, was the center of a bustling fur trade between North America and China. Britain and Spain both laid claim to trading and, increasingly, settlement rights on the island, leading to a dispute that threatened to erupt into war. George Vancouver, the captain of the Royal

Navy ship *Discovery,* traveled to the outpost to try to resolve the differences between the two nations.[12]

Vancouver had previously visited Nootka Sound with Captain James Cook in 1778. Although Vancouver failed to reach an agreement with Spain, he had additional orders to fill in the gaps in Cook's previous mapping of the Northwest coast. On leaving Nootka, Vancouver steered his ship southeast, seeking an inside passage mapped by the American captain Robert Gray a few years earlier. He sailed through the Strait of Juan de Fuca and turned south into a great inland sea—the Whulge.[13]

Vancouver, then thirty-five years old, kept a journal of his explorations that was published after his death just five years later. He named these waters Puget's Sound for his lieutenant, Peter Puget. His journal describes exchanges with the local Native people, along with descriptions of their homes, foods, and customs. He notes that many homes had been abandoned, with human remains visible along the shorelines, and that signs of warfare (trophy heads) and Western disease (smallpox scars) were common in the settlements they visited.[14]

On May 19, 1792, Vancouver anchored off a promontory called Sc'utqs, now known as Restoration Point on Bainbridge Island, to make repairs to his ship. The six-year-old Se'alth and his family watched the ship ride in on the tide and heave to a stop in the deep water just offshore of their harvest camp. The ship, which dwarfed the largest of his village's canoes, carried one hundred men. Stories handed down through the generations describe a behemoth that came from the north, casting its shadow over the tide flats. In some accounts it is said to have been taken for an island that had broken loose, advancing across the water toward them, or for a giant mythical bird. Yet according to Vancouver's journal, Se'alth's family and other villagers seemed relatively unalarmed by their arrival. He wrote that "several natives assembled to view the ship as we passed by, and two paddled a canoe around the anchored ship before returning to shore."[15]

And that was it. There was no more interaction between Vancouver's crew and the members of the camp that day. The following morning, Vancouver noted, "In the meadow and about the village, many of the natives were seen moving about, whose curiosity seemed little excited on our account."[16] Vancouver's curiosity, by contrast, got the better of him. He went ashore, recorded the latitude, and visited the camp, reporting that "nearly the whole of the inhabitants belonging to the village, which consisted of about eighty or an hundred men, women and children, were busily engaged like swine, rooting up this beautiful verdant meadow in quest of a species of wild onion, and two other roots."[17]

The village was not a "miserable" permanent settlement, as some white observers assumed, but a seasonal Suquamish camp called Tactu. The inhabitants were likely harvesting lily bulbs and native hyacinth, or camas—dietary staples cultivated throughout the Whulge coastal area. They invited Vancouver's party to share a meal of shellfish they had collected from the beach.

Vancouver describes two of the group's leaders as "particularly assiduous to please." The men gestured to indicate that they would visit the captain on his ship later that day. The group later approached the ship singing and beating a rhythm on their canoes with their paddles in ways similar to the ceremonial traditions Vancouver had witnessed in Nootka Sound. "This performance took place whilst they were paddling slowly round the ship, and on its being concluded, they came alongside with the greatest confidence, and without fear or suspicion immediately entered into a commercial intercourse with our people." Vancouver noted that the visitors used iron-tipped arrows, which they were eager to trade away, and wore copper jewelry. When the two men he identified as chiefs boarded the *Discovery*, they were interested in acquiring copper and little else.[18]

Over the next couple of days, all of the encampment's men, women, and children paddled out to trade with the ship, though no others came aboard. A few of the villagers paddled across the sound to notify their relatives and neighbors on the mainland—the Doo-Ahbsh residents of the area comprising current-day Seattle—who returned with them to join the trade with the ship's crew.[19]

Although the *Discovery* may have been the first ship to venture that far into Puget Sound (or at least to record its visit), its arrival, as Vancouver's description makes clear, was not the first contact between the region's people and European traders. In fact, local peoples' trade routes extended north to Vancouver Island, including Nootka Sound, and east beyond the Rocky Mountain Range. Montana's Blackfoot Indians share linguistic and cultural roots with Puget Sound's tribes, some of whom may have migrated inland during the early scourge of smallpox that hit the Whulge prior to Vancouver's first visit.[20]

Se'alth's father, Schwabe, and his uncle, Kitsap, may have been the two "chiefs" who boarded the *Discovery* that day, though Vancouver never apparently asked or recorded their names. During the following days, while the ship sat at anchor, the Suquamish leaders and their villages made extensive trade deals for Western goods that elevated their wealth and status in the region. The ship's visit made a significant impression on young Se'alth, who often talked about it later in life. The trade relations established on that trip likely influenced his dealings with later European traders and, eventually, with the first American immigrants to arrive on the shores of the Whulge.[21]

A CONFEDERACY OF TRIBES

In the years following Vancouver's 1792 visit, Puget Sound became an important stop for British, Spanish, and Russian traders sailing between North America, Europe, and China. Pulling the Pacific Northwest tribes into the global economy, traders exchanged textiles and other manufactured goods for animal skins harvested by Native hunters. Otter pelts were in especially high demand and were shipped by the thousands to wealthy elites in Chinese port cities. Beaver, elk, and bear skins were also traded for flour, sugar, metal tools, and guns. Raised in this budding market economy, Se'alth incorporated the lessons he learned from it when American pioneers began settling permanently in the lands around the Whulge.[22]

Shortly following Vancouver's visit, and supported in part by the wealth acquired from trading with the *Discovery*, Kitsap built a new longhouse for his people. Tsu-Cub, or Old Man House, became a prominent village at Port Madison, just north of Bainbridge Island. In time, it became the gathering place of six allied tribes—a central Puget Sound confederacy that sought to defend itself against a growing number of raids by rival tribes to the north and south.[23]

Se'alth's uncle Kitsap was a noted warrior, while his father was a respected peacetime leader. Se'alth too grew up to become a respected

leader among the Suquamish. Family lore suggests that he initially emu-
lated the more combative role of his uncle. His influence increased during
the 1820s, after he defeated a planned attack on the lowland saltwater tribes
by raiders from the mountainous areas of the White River, in the upper
Duwamish watershed.[24]

The best account of his triumph comes from Samuel Coombs, a settler
who arrived in Puget Sound in 1860. In 1893, he reminisced about his first
encounter with Se'alth in an interview with the *Seattle Post-Intelligencer*.
He recalled happening on a council meeting on the beach fronting Seattle
and noted the respect and deference other Native leaders showed Se'alth.
Using an interpreter, Coombs interviewed several local headmen about
how Se'alth had earned his venerated position.[25]

Having heard about the southern raid in advance, perhaps from upriver
relatives of Se'alth's Duwamish mother, the allied saltwater tribes gathered
at Old Man House to plan their defense. They decided on an ambush
strategy proposed by Se'alth. According to Coombs, the defenders selected
a sharp river bend above the Black River, where the river flowed swiftly,
and felled a fir tree just above water level in order to capsize the advancing
canoes. They then hid at the top of a neighboring hill. "As soon as it was
dusk, five large canoes loaded with one hundred selected warriors started
down the stream, and as there was a strong current it was not long before
it fell into the trap," Coombs recounted.[26]

Thirty of the raiders were drowned, killed, or captured, and the rest fled
on foot back upriver. "When Sealth and his warriors returned to the bay
with such substantial proofs of the victory gained over their former perse-
cutors great was the rejoicing among the saltwater tribes and the hero of
the hour was the young warrior who, by his cleverness, boldness and courage, had delivered them from a great danger," reported Coombs. As a result,
Se'alth's star quickly rose.[27]

According to early settlers who came to know Se'alth and heard the
stories of his military prowess, he "soon overshadowed all the chiefs in that
region and became their recognized leader." While perhaps not all of the
region's tribes would agree, the Duwamish Tribe's historical records report
that the six confederated tribes of Old Man House selected Se'alth to
advise them. The confederated tribes included the lower Skagit from

Whidbey Island, led by the elder Chief Kanum and his son, Kruss Kanum—Tupt-Aleut's husband, later known as George Snatelum. Old Man House grew in size and prominence under Se'alth's leadership.[28]

To understand how Se'alth, Kanum, and their allies dealt with the American immigrants to Puget Sound, it is helpful to understand Northwest family and village ties at the time. The concepts of tribes and chiefs were introduced by the settlers and did not reflect the social organization of Puget Sound's people. Leadership was largely local, complemented by family alliances. The people who lived in the greater Seattle area at the time were part of an interconnected network of autonomous villages stretching from the Duwamish River delta near modern Seattle's downtown up along the Green, White, Black, and Cedar Rivers (where the peoples of individual villages all had their own names, including Doo-Ahbsh, Skwohp-Ahbsh, and Stook-Ahbsh), and throughout communities on Lake Washington (referred to as Hah-chu-Ahbsh, "lake people"), Lake Union (Ha-ha-chu-Ahbsh, "littlest lake people"), Salmon Bay (Shill-shohl-Ahbsh), and Green Lake.[29]

While *Doo-Ahbsh* is often translated as "people of the inside" and is now used to refer to all of the Duwamish watershed's inhabitants, the early-twentieth-century ethnographer T. T. Waterman traces the name to a single village on a tributary of the Black River known as "inside place." Each community had its own unique name to identify itself. Cultural and family ties connected the villages, but they did not have a centralized regional system of governance.[30]

Families and villages typically formed trade and political alliances with other Native groups in the region, many of which were reinforced by marriage. Despite these alliances, intertribal rivalries, raids and outright warfare also occurred, as demonstrated by the defeated raid that led to Se'alth's rise. Village leaders, or *sia'p*s, were part of an upper class, guiding and protecting their extended families and allied villages. High-status families, like those of Se'alth, Tupt-Aleut, and Kanum, often kept slaves, usually captives from raids on rival groups. Political and economic power were interrelated. A leader's stature and influence were conveyed by his wealth—not by how much he accumulated and kept, but by how much he gave away during tribal celebrations and gatherings.[31]

To make matters more confounding from the settlers' perspective, Northwest Native people did not identify as members of just one group. Se'alth was Suquamish on his father's side and Duwamish on his mother's side. And both his parents had ties that further connected the family to other groups in the region. The descendants of Tupt-Aleut and Kruss Kanum claim an extensive family network that includes people now identified as Duwamish, Suquamish, Snoqualmie, and Swinomish, as well as Yakama, far over the mountains. The larger trade networks facilitated by these ties extended even farther, north to Vancouver Island and as far east as Montana.[32]

This was the context into which Vancouver sailed with his crew aboard the *HMS Discovery* in 1792.

TIES THAT BIND

James Rasmussen stood at the precipice, peering down. It was 1986, a decade after the discovery of the Duwamish fishing village at the Port of Seattle's Terminal 107. Another discovery had recently halted construction of a new apartment building and shopping center in Renton. Rasmussen, then the newest member of the Duwamish Tribal Council, went to inspect the site. He stood above an enormous construction pit, its excavators standing idle. The road beneath his feet followed the historical riverbed that once carried the waters of Lake Washington through the Black River to the Duwamish and on to Puget Sound. In the pit below him lay the remains of Sbabadid, or Little Hills, his maternal family's ancestral village. "My history is going to be buried," Rasmussen lamented after being told that nothing could be done to stop the construction. "That means it's gone forever."[33]

Six feet tall, Rasmussen carried himself with pride. His brown hair was pulled back in a long ponytail, and he sported a thick beard. He was a notable musician in Seattle's jazz scene, and at age twenty-four, was following in the footsteps of his mother and grandfather on the tribal council. His great-great-great grandmother was Tupt-Aleut, who was born here when Se'alth was still a young man and before any white immigrants had settled in the region. According to the family's oral history, the residents of Sbabadid were high-born, among the leadership class of their tribe.

As an adult, Tupt-Aleut lived in the area now known as Coupeville, on central Whidbey Island, with her husband, the Skagit headman Kruss Kanum. They traded with the English, and they welcomed the arrival of Colonel Isaac Ebey, the first American to successfully settle on the island in the early 1850s.[34]

Kruss Kanum and Tupt-Aleut became known to the new settlers as Chief George Snatelum and Tyee Mary. They had two daughters: Quio-litza, later known as Ann, and Ellen, whose Native name has been lost to history. Kruss Kanum also had children from previous wives. The girls' half-brother, George Jr., and their father signed the treaty that created a system of reservations for the Puget Sound tribes. George Jr.'s son, Charlie Snatelum, was lauded at his death in 1934 as the last of the "Indian Chiefs" of Whidbey Island.[35]

As young women, Quio-litza and Ellen returned to their mother's village of Sbabadid, where they rejoined their mother's brother, a noted shaman named Sdabahld. According to family history, Quio-litza was then captured in a raid by members of the Yakama Tribe of the eastern plains. Their oral history tells of her abduction and dramatic return: she is said to have crept away from her captors under cover of night, stealing a canoe and paddle to escape by river, and then finding her way through the mountains back home to Sbabadid. The stolen paddle, carved of cedar, is a treasured keepsake still in the family's possession.[36]

There is some evidence in the family's genealogical records that Quio-litza's mother had family ties to the Yakama, adding to the intrigue of the story. Was Quio-litza abducted, or did she elope? Did she escape imprisonment or slavery, or was she running away from a bad marriage? Quio-litza's great-granddaughter, Ann, recounted the story of the raid in 1998. "This has been relayed to me by my father, and he was not one to tell fairy stories or tall tales," she said. "He told me that she got away and she came down the rivers."[37]

Quio-litza's daughter later applied for an allotment of land on the Yakama reservation for her son, again raising the question of whether there were family ties. To Ann Rasmussen, however, her daughter's demand was sufficiently justified by the abduction. "I think she was really trying to get even with the Yakamas for what they did to her mother," said Rasmussen.

Ann Rasmussen, great-granddaughter of Quio-litza (aka Ann Kanum), holding the family's heirloom canoe paddle. According to her Duwamish family's oral history, Quio-litza used the paddle to escape from the Yakama Tribe after being kidnapped. Courtesy of Jeff Corwin.

The story reflects the complex ties that bind many tribes and adds intrigue to later alliances between the Duwamish and Yakama tribes that would play out dramatically in the years following white settlement.[38]

FIRST AMERICANS

During Quio-litza's teenage years, parts of Puget Sound were beginning to sprout small settler communities. For decades, Americans and British had coexisted under the terms of a treaty between the two nations that allowed them to trade but not to establish permanent settlements. In 1840, the British-owned Hudson Bay Company established a trading post in southern Puget Sound near the Nisqually River. They built trade relationships with Native groups and exported pelts and other goods. The company's records painstakingly detailed trades, labor payments, and exchanges with Native and foreign traders, as well as with early Americans exploring the region: the number of beaver pelts and other furs bought and sold, the prices paid to Native traders, and the schedule of shipments to China and other ports.[39]

Company officials also kept journals describing major events and countless everyday interactions at or near the trading post. Buried in these records are notes on the visits of Se'alth, Kanum, Leschi of the Nisqually, and other Native traders, laborers, and visiting dignitaries. David Buerge summarizes Se'alth's dealings at the company store in his biography *Chief Seattle and the Town That Took His Name*. Se'alth's interactions with the British traders were often tense, if not outright hostile. In the mid-1830s, the company's records describe Se'alth as argumentative and violent. In one instance, he drunkenly fought with the Nisqually headman Lahalet (Leschi's father). Se'alth later threatened Jean-Baptiste Ouvre, the interpreter who tried to break up the fight, at gunpoint.[40]

Leschi and Kanum, in contrast, were trusted employees of the British company. Kanum worked as a courier, while Leschi and his brother tended the company's horses. Despite the company's practice of lowering prices paid for goods once trading relationships were established, its staff stayed on good terms with the Nisqually people, and several Hudson Bay Company officials married Native women during their years in Puget Sound.[41]

American settlers began arriving in the region in the mid-1840s. Among the earliest arrivals was Michael Simmons, who would play a key role in US–Native relations as a federal Indian agent in the years ahead. The new immigrants found themselves in a society made up of British trade officials and Puget Sound's Native communities, along with a transient assortment of European, Asian, Pacific Islander, and African seamen from the ships transporting furs along the Pacific trade routes. Two decades of mostly cordial and prosperous trade relationships between the British and the local tribes were soon dramatically altered by the more permanent aspirations of the new arrivals.

The new influx was spurred by the 1846 Oregon Treaty between the British and American governments. The treaty settled a long-standing border dispute by dividing the Northwest between the two nations at the forty-ninth parallel of latitude, which later became the US-Canadian border. With the territory secured for America, the US Congress passed the Donation Land Claims Act in 1850, offering free land to any of its (white male) citizens who agreed to settle on the new frontier. Under the act, 160 acres of land would be given to any settler who undertook to "improve" his property. Those who had arrived before 1850 were eligible for 320 acres, and any married male settler could claim an additional allotment for his wife, doubling their land holdings.[42]

Luther Collins and his family had arrived in 1845, settling in the Nisqually prairie near the Hudson Bay Company. The company's records provide scattered insights into their lives there. For several years, Collins farmed, worked occasionally as a day laborer for the company, and disappeared for long stretches to join the hordes drawn to the California gold rush.[43]

On his second trip to the goldfields in 1851, Collins met several other Western pioneers, with whom he would set out on a lifelong adventure. They soon found themselves in the center of Se'alth's world, at the beginning of the story of the city that now bears his name.

2

And Then There Was Blood

Newcomers, Alliances, and Betrayals

1850–1900

IN THE FALL OF 1865, MARY ANN CAVANAUGH BEGAN MAKING PLANS for a winter feast at her family's homestead on the Duwamish River. The days were shortening, and the past few years had taught her to anticipate the isolation and loneliness of pioneer winters in the Northwest. She and her family drew up a guest list, commandeered post carriers and canoes to deliver the invitations, and foraged for blackberries and Oregon grape near the river. When the fall floods swamped the valley, her family and neighbors harvested potatoes and flushed ducks from the reeds, and as winter approached they hunted for venison and bear. On the morning of January 16, 1866, Mary Ann rose early and began making her final preparations.[1]

Her family had settled on the Duwamish River in September 1851. The US government's promise of 160 free acres to every settler who put down roots and "improved" the land had drawn her father and brother north from the California goldfields. Their land claims were the first in this part of the Oregon Territory, just upriver from a deep embayment in Puget Sound. Mary Ann's father, Jacob Maple, and her brother, Samuel, had been persuaded to survey for land along the river by the Nisqually settler Luther

Collins, whom they had met when the three were returning from the goldfields. When they arrived, the river delta and tidal estuary snaking south of it hosted fifteen longhouses dispersed among five villages. In the network of rivers and lakes making up the larger watershed, at least ninety longhouses housed an estimated ten thousand Native people.[2]

Collins's original land claim was about fifty miles south of the Duwamish Valley, and he had visited the fertile river basin before. He knew that white settlers could anticipate the support of a local Native leader, Se'alth, who had been recruiting pioneers—unsuccessfully, to date—to settle in his territory and form alliances. Se'alth had provided them with a team of Native guides for a scouting visit in June 1851. Now, nearly fifteen years later, Se'alth was expected as a prominent and honored guest at the Maple-Cavanaugh clan's winter feast.[3]

Luther Collins's family and the Dutch immigrant Henry van Asselt, who settled on the river along with Jacob and Samuel Maple in 1851, had been their closest neighbors ever since. Another brother, Eli Maple, arrived in 1852. A decade after their father, the remaining Maple children made the six-month trip to the frontier from their home in Ohio. Mary Ann arrived with her husband, Martin Cavanaugh, and their two-year-old daughter in November 1862. Four years later, as she prepared for the winter feast, Mary Ann was pregnant with their second child.[4]

In the years after Jacob and Samuel Maple began clearing their new land, other pioneers established farms and settlements near the family's homestead. Settlers moved up the Duwamish River to its tributaries, the Black, Green, and White Rivers. When the Maple siblings arrived, they cut roads, logged, mined, and farmed land along the Duwamish River and its tributaries. The town of Seattle, where the Denny Party settled on the bay, had been incorporated in January 1865. Guests from all of these settlements traveled by foot and canoe to join the feast at the Maple-Cavanaugh house. Henry Yesler came in his horse-drawn wagon, traveling via a rough trail connecting the Duwamish River to Seattle along the base of Beacon Hill. Other members of the Maple family traveled from Oregon and over the Cascade Mountains to join in the festivities. It was the social event of the year.[5]

Gene Gentry McMahon, *The Twins Patrol the River* (2017). Ink wash on paper, 16 × 12 in. Artist's depiction of Cora and Dora Maple, granddaughters of the Duwamish River pioneer settler Jacob Maple.

The fledgling city at the mouth of the river had been incorporated under the anglicized name of the tribal leader who was the Maple and Denny parties' patron in their early years. Se'alth welcomed and supported the early pioneer families, seeing to it that they survived the first years in their new surroundings. Se'alth recruited some of these transplants to bring trade, protection, and manufacturing to his villages. He was both a business partner and benefactor to many of these early American families. By the 1860s, Se'alth had earned the trust of many of the pioneers, and although the favor was often not returned, he remained their ally. To some, including the Maples, he was also a lifelong friend. At the time of the 1866 feast, Se'alth was nearly eighty years old.[6]

Most of the guests were already present when Se'alth arrived at the party. Rounding the river bend in a massive canoe propelled by fifty paddlers, his entourage slapped the water and sang a traditional welcome song as they approached. As Se'alth came ashore, Mary Ann recalled the chief stepping from his canoe "with all the dignity of his rank" and formally greeting the assembled settlers. His retinue then unloaded gifts of cougar skins, moccasins, fish, and salmon roe—an expression of Se'alth's friendship as well as his status and wealth. The bill of fare for the feast reflected a panoply of wild foods introduced to the pioneers by Se'alth's people:

SOUPS: Chicken Soup, Clam Chowder, Stewed Clams
FISH: Smoked, Boiled, and Fried Salmon, Salmon Trout
FOWL: Grouse, Duck, Quail, Snipe
ROASTS: Venison, Smoked Bear, Ham
VEGETABLES: Mashed Potatoes, Roasted Squash
CONDIMENTS: Honey-in-comb, Preserved Wild Blackberries,
    Oregon Grape Jam

The guests were seated on makeshift benches the family had built from cedar boards and tucked into every available corner, and dinner was followed by toasts and stories that extended late into the night. Many of the settlers' tales of their earliest years together acknowledged the tremendous help they had received from Se'alth, but they also recounted the many clashes they had had with "hostile" members of his and other nearby tribes.[7]

It was the last time Se'alth would visit the settlers in the city they had named for him. Six months later, he passed away at his home on the Suquamish Tribe's Port Madison Indian Reservation, the location of his family's expansive longhouse, Old Man House. Port Madison, on the Kitsap Peninsula across Puget Sound, was the only land reserved for Se'alth's people following the treaty he had been persuaded, or obliged, to sign by the territory's colonial governor a decade earlier. No reservation had been created within his maternal Duwamish family's ancestral lands, where Mary Ann and her family lived, and no newspaper reported his passing.[8]

AN UNEASY ALLIANCE

Se'alth's relationship with the Maples and their contemporaries was the product of his decision to launch a grand experiment: building a multi-ethnic society in his homeland, extending from the upper Duwamish River village of his Doo-Ahbsh (Duwamish) mother to the territory across Puget Sound formerly governed by his Suqw-Ahbsh (Suquamish) father. Often in opposition to members of his own extended family, he sought to coexist with settlers in the Puget Sound region, intending to turn their arrival to his advantage and that of his people. After careful observation, he sought to ally with the new American settlers, who were arriving in increasing numbers at the southern edge of the Whulge, or Puget Sound.[9]

Se'alth was not alone in pondering how best to manage the new arrivals. The Snoqualmie leader Patkanim had not been receptive to the first Americans who sought to settle in the region. In 1848, while Luther Collins was still farming near Nisqually and the settlers numbered only a few dozen, Patkanim had advocated that the central Puget Sound tribes combine forces to expel them.[10] Then in 1850 he traveled with one of the newly arrived traders by ship to San Francisco to see where the settlers were taking all the trees that had been felled from his homeland's forests.

Patkanim returned with a description of the bustling, developed city and the thousands of Americans filling its streets—city dwellers who had begun to move north in search of new opportunity and new land, and who far outnumbered the combined population of all of their Puget Sound tribes. The region's Native communities coveted the Americans' trade goods and technologies and saw that they could be either powerful allies or formidable foes. Patkanim switched allegiances and became an ally of the new settlers, along with Se'alth and several other Puget Sound leaders. It was part of a calculated strategy to survive, and perhaps thrive, as the number of settlers in their territory began to swell.[11]

In 1850, a scouting party of Americans traveling north through Puget Sound encountered Se'alth and members of his village fishing in the bay at the mouth of the Duwamish River. Two large villages of three to four longhouses each—one on either side of the river delta—would have been visible to the party as they entered the bay. One of the group, Colonel Isaac Ebey, was in search of land to put down roots, and the group stopped to speak with the Native fishermen. The Hudson Bay Company employee B. F. Shaw described the landing of their canoe on a beach "alive with moving Indians." They were greeted with "such a whooping, such a jumping stiff-legged, such a shaking of knives and blankets and shooting off of guns, I had never seen before." Then "a large, middle-aged Indian, with a very wide head" walked out on the logs jutting from the shore and spoke to the settlers.[12]

"My name is Se'alth, and this great swarm of people that you see here are my people," the imposing man began. "They have come down here to celebrate the coming of the first run of good salmon. . . . This is the reason our hearts are glad today, and so you do not want to take this wild demonstration as warlike. It is meant in the nature of a salute in imitation of the Hudson Bay Company's salute to their [British] chiefs when they arrive at Victoria."[13]

Se'alth welcomed the travelers, inviting them to bring guns, axes, clothing and tobacco, flour, and sugar. He didn't have to ask twice. "We have come on a mission of peace and friendship, and as our people are becoming so crowded in the East we are looking for a place to locate them," Shaw said. He promised that the settlers would build sawmills, flour mills, sugar mills, blanket mills, and stores "where everything will be kept for trade with the Indians." He also promised to provide schools to teach Native children "to do all the things that we can do."[14]

After their exchange, Shaw's party continued north, and Ebey selected a piece of land near the Skagit leader Kruss Kanum's village on Whidbey Island. Shaw returned south to the trading post at Fort Nisqually. Their story of the encounter with Se'alth no doubt made it to Olympia, where a settlement of rough homes, merchant shops, and crude manufacturing had already grown up on the south shore of Puget Sound. Se'alth himself did not wait for word to travel through the settlers. In the following months, he traveled to Olympia himself and began recruiting partners to establish business enterprises and residences in his territory.[15]

On one of these trips, Se'alth met Captain Robert Fay, a shipping merchant with whom he struck a deal to catch, preserve, and export the season's abundant salmon. In 1851 Fay took their first shipment of native-caught salmon to San Francisco, but he did not immediately settle in the Seattle area. Se'alth returned to Olympia to recruit more Americans, returning with "Doc" Maynard in 1852. Together they caught and packed more barrels of salted salmon, shipping them off that fall with a load of timber, but most of the salmon spoiled en route. Maynard abandoned the salmon trade, but he stayed in Seattle and maintained his business dealings with Se'alth. Their partnership would last until Se'alth's death.[16]

In September 1851, Se'alth launched his effort to form a cross-cultural society on the banks of the Duwamish River with Jacob and Samuel Maple, Diana and Luther Collins and their children, and Henry van Asselt. His decision to encourage American settlement changed the course of history for the Duwamish River, the Duwamish Tribe, and settlers in and around the future city of Seattle. Se'alth was gambling on changing the history of his people for the better, rather than for the worse, given the numbers and technologies the settlers brought with them.

Shortly after the Maples and their companions arrived, scouts for the Denny Party landed at the southwest corner of Elliott Bay and built a simple shelter near the Duwamish village of Tualtwx, or Herring's House. The rest of the Denny Party arrived with Captain Fay from Olympia in November 1851, when Fay stopped to collect the salmon Se'alth's fishing crews had caught for him.[17]

Within a year, most of the group relocated to a narrow shoreline on the east side of the bay, tucked between a ridge of high bluffs and a deepwater port that was better positioned to receive the large sailing ships that were beginning to frequent Puget Sound. This new settlement was christened Seattle and was formally incorporated as a township of the new Washington Territory on January 14, 1865.[18]

Se'alth's tribe and several others regularly visited to trade with the new arrivals or had managed to recruit and form trade relationships with other settlers near their own villages. At least a day's paddle from the Hudson Bay Company's Nisqually trading post, and farther still from the merchants at Olympia, the new settlements provided the central Puget Sound tribes with trading and business partners much closer to home.

Farther north, Patkanim of the Snoqualmie Tribe, around present-day Everett, and Kruss Kanum, the Skagit leader on Whidbey Island, had begun to receive settlers in their own territories and to forge trade and defense alliances with the newcomers. But the Duwamish River and Elliott Bay easily had the largest concentration of homesteaders during the early years of settlement. The new immigrants quickly set up local businesses that catered to and employed Se'alth and his allied Duwamish and Suquamish villagers in addition to their few fellow settlers.

During their first winters, having arrived with limited supplies, the settlers struggled to adapt to their new environment, relying heavily on the knowledge and assistance of their Duwamish hosts. The homesteaders watched as tribal fishermen caught and dried the river's returning salmon, harvested native plants, and hunted for wild elk and winter waterfowl. The settlers learned that the land was rich with wild and cultivated foods and materials that could provide winter clothing, blankets, shelter, firewood, and ceremonial supplies.[19]

The settlers' first winter was lean, but they were few in number and were able to trade for necessities. In the spring, after surviving a winter of dependence, the Denny Party made arrangements for one hundred bushels of potatoes to be delivered from the Hudson Bay Company at Nisqually. Edward Huggins made the delivery in the company's large mail canoe, which was on its way to Vancouver Island. The canoe dropped Huggins and the potatoes at the nascent settlement, and Huggins scouted about for a return ride.[20]

He found Wyamook and his wife, a Nisqually couple he knew from the trading post, visiting with Se'alth at Alki Point. Wyamook agreed to take Huggins back with him. Huggins found the trip harrowing in the canoe "so small in size that a strong man could carry it upon his shoulders." Even with better transportation, traveling so far to trade with the small central Puget Sound settlements was not sustainable. By the end of the year, Terry Lowe, a Denny Party settler, and the newcomer Doc Maynard would both open shops stocked by shipments from Olympia and San Francisco, greatly reducing the settlement's reliance on the Nisqually trading post.[21]

Even so, Arthur Denny later recalled that the second winter was even more difficult than the first. Despite having had more time to plant and store crops, the settlers were not prepared for the infrequency of winter supply shipments. "Few vessels visited the Sound for several months, and as consequence it was a time of great scarcity, amounting almost to distress," wrote Denny. The settlers ran out of flour and bread, but, "with fish and venison we got along quite well while we had potatoes. Finally they gave out." Their Duwamish hosts were away at their winter villages on the Black River, so a group of settlers traveled more than twenty miles upriver to trade for a bushel of "wapatoes"—a local relative of the potato. The tubers of a native marsh plant, they were harvested primarily by Duwamish women, who dug them out of the shallows with their toes.[22]

In the meantime, the Duwamish River settlers developed their farms to provide for their own needs. In 1851, the Collins family had planted fruit trees along the river, using seed they had brought with them. Introduced trees, planted in uniform orchard rows, began to replace the native cedar, spruce, and riparian assortment of alder, willow, and cottonwood. The new crops quickly began to stamp the settlers' mark on the landscape. The trees grew well in the fertile river valley, and by 1855 the Collins orchard boasted over one thousand fruit trees and produced three hundred bushels of peaches for sale.[23]

When Eli Maple joined his father and brother in 1852, he went to work on Collins's farm, then began harvesting timber for the mill built by the new immigrant Henry Yesler. With San Francisco clamoring for lumber to rebuild the city after its great fire in 1851, Jacob and Eli felled trees in the Duwamish Valley by the thousands, dramatically altering the local landscape. Eli later recalled, "My father and I took a contract for getting out 7,000 telegraph poles and 5,000 boat pols [sic]; these we packed out of the woods to the water on our shoulders." Denny's daughter would later reflect, "The most astonishing change wrought in the aspect of nature by the building of a city on Puget Sound is not the city itself, but the destruction of the primeval forest."[24]

The Maple father and son earned enough from their harvests of raw timber to buy a team of oxen, transporting them from the Columbia to the Duwamish River. "There we went to farming as well as lumbering," recalled Eli Maple in his journal. All along the Duwamish River, settlers' farms, filled with introduced crops and grazing livestock, were replacing groves of native trees and expansive prairies.[25]

His orchards well established, Luther Collins turned to new ventures that would continue to alter the landscape and bring business to the region. He founded the first ferry on the Duwamish River in July 1853. In October of that year he helped found the Duwamish Coal Company on the Black

River with the newly arrived physician and entrepreneur Dr. Ruben Miles (R. M.) Bigelow. The fledgling company extracted and shipped three hundred tons of coal down the river the following summer.[26]

It was Bigelow who first discovered coal deposits near the confluence of the Cedar and Black Rivers, where the enormous inland watershed of Lake Washington drained into the Black River. This was the location of Sbabadid, where James Rasmussen's maternal ancestor, Tupt-Aleut, lived before marrying Kruss Kanum. After the discovery of coal, the lands around Sbabadid became some of the most valuable property in the region, drawing more settlers—speculators and laborers alike—to the area they called Lake Fork.[27]

By the time the Fork's first settlers arrived, Kruss Kanum and Tupt-Aleut's daughters, Quio-litza and Ellen, were of marriageable age and had returned to the land around Sbabadid. Their lives soon became integrated with those of the early Black River settlers. Duwamish women had begun to marry the new immigrants as they traditionally had intermarried with other tribes, securing their support and gaining access to new resources and trade relationships.

Tribal records show that Quio-litza married Bigelow, a widower with no children, around 1853. His status as a doctor and his coal fortune would have made him an attractive ally for the families at Sbabadid. Bigelow and Quio-litza had a daughter, Katie, in 1856 and a son, Miles Jr., three years later.[28]

Despite Bigelow's early success, his Duwamish Coal Company soon began to fail as a result of poor financial management and the disruption of the Indian Wars. It spawned successors, however, and coal mining became one of the dominant industries in the Duwamish watershed, particularly in the area now encompassed by the city of Renton at Lake Fork, as well as in Newcastle and the aptly named Black Diamond.[29]

Other businesses also thrived. A sawmill built at the confluence of the Black and Cedar Rivers provided fuel to the coal factory and employed Duwamish laborers, at least some of whom seemed to take pride in the bustling new businesses. "What afforded the greatest satisfaction to [the Duwamish] was its situation upon their property, and the superior importance thereby derived to themselves," wrote the historian Clarence Bagley. "They . . . took every visitor through the building to explain its working, and boast of it, as if it had been of their own construction."[30]

As settlements extended along the Black and Cedar Rivers, the Duwamish River became a food production center and a transport route for logs and coal. In March 1853, the *Columbian* newspaper proclaimed, "Puget Sound is naturally intended and must eventually be the great commercial mart of the Pacific Coast, for the fishery, lumber and coal trade." The early settlers fully embraced their success as confirmation of this manifest destiny.[31]

When the Elliott Bay settlers decided to name their town for their Native patron as a mark of their great respect, the idea was reportedly anathema to the aging Se'alth. It violated Native traditions prohibiting the names of the deceased to be spoken. Se'alth's daughter, Angeline, told the *Post-Intelligencer* in 1891 that she traveled to Olympia with her father to persuade the colonial governor to intervene, but his pleas were ignored.[32]

All through this time there were occasional conflicts, but few were violent. The pioneers relied heavily on the knowledge, labor, and friendship of the local tribes. By 1854, however, this honeymoon period was fast drawing to a close.

THE INDIAN WARS ON DUWAMISH LAND

Despite the accommodating stance and interdependence of the settlers and the Native communities on Puget Sound, not all individuals, on either side, took well to sharing the territory. The settlers brought a system of private property ownership that conflicted with generations of traditional resource use, and the Native people were keenly aware of the immigrants' growing numbers. They knew, too, that many more were likely to arrive. For some, that was a reason to build alliances; for others, it was reason to keep their distance or try to keep the newcomers from settling in their territories altogether.

The first deadly conflict between Natives and settlers in the Duwamish colony occurred on Luther Collins's farm. As an early arrival to the territory and a married man, he qualified for a land claim of 640 acres—the largest held by any of the original Duwamish River settlers. His farm covered most of Seattle's modern-day neighborhood of Georgetown and part of Beacon Hill. It encompassed a Duwamish village with two long-houses known as Place of the Fish Spear and included a large prairie, cultivated by years of prescribed burning by the river's Native people.[33]

The Duwamish Valley's cleared prairies, which increased the diversity of plants and wildlife available for harvesting and hunting, were part of what attracted settlers to the river. The Duwamish River had been described in detail by Colonel Isaac Ebey during his visit while scouting for a home-stead in 1850: "The river meanders through rich bottom land, not heavily timbered, with here and there a beautiful plain of unrivaled fertility." The land was highly valued by the river's settlers and Natives alike. After Col-lins filed his claim, Native people continued to use the land to hunt and harvest local plants, while Collins began to plant his own crops, both to feed his family and for sale.[34]

In 1853, Collins killed a Native man named Masachie Jim, who was suspected of beating and murdering his own wife. Masachie Jim may have been harvesting camas in the prairie or simply returning to his village when Collins and two other settlers spotted him, captured him, and lynched him. Collins exercised vigilante justice again the following year in retaliation for the murder of an American settler visiting from Olympia. Two Native men suspected of the crime were being held for questioning in a cabin. Collins and a small group of men broke into the cabin and hanged them. Collins and his collaborators were in turn charged with murder, but several members of the jury in the small settler community doubled as witnesses for the defense. The trial was a farce. Collins's codefendants were declared not guilty, and the charges against Collins were dropped. Later that year, Collins was elected King County commissioner.[35]

Luther Collins was not alone in assuming the right to dole out punish-ment to Native people, but his violent acts certainly destabilized the relation-ship that the early settlers and their Native hosts had hoped to establish.

The earliest land claims filed by the Maples and other Americans were complicated by the fact that the US government had not yet acquired legal rights, under its own laws, to the lands that were being settled. To do so, they needed to secure treaties with the Indigenous people who already occupied the land. While treaties have often been used as a means of ending a war, the collection of treaties wrested from the Native tribes in the Pacific Northwest sparked one instead. Rushed and coercive negotiations that offered American goods and services in exchange for Native land met with increased objections at each successive treaty signing pushed by the new territorial governor, Colonel Isaac Stevens.[36]

In the winter of 1854–55, Stevens embarked on a whirlwind series of treaty conventions with dozens of Native communities. He imposed a structure to streamline the process, assigning each village to a "tribe" represented by appointed "chiefs." Though the settlers knew that the Native people did not have chiefs, the governor's ethnographer advised that "the wisest course for the government to pursue seems to be to aggrandize a few principal chiefs at the expense of the petty tyees; to recognize the former alone and hold them responsible for all acts committed by their people. They could thus be compelled to exercise an authority which they did not before possess." The independent ethnographer T. T. Waterman noted several decades later: "The men called 'chiefs' by the whites had a largely fictitious authority, excepting, of course, as they were inducted into office by a blundering government for purposes of treaty-making and treaty mongering."[37]

Stevens left it to his designated "chiefs" to identify and bring additional village leaders to the table, as "subchiefs," to join in the treaties. He appointed Se'alth to represent all the Duwamish and Suquamish people living in the Duwamish watershed, on Bainbridge Island, and throughout the peninsula to the west. Despite—or perhaps because of—his lack of authority over the diverse group of freshwater and saltwater villages in this vast area, Se'alth duly selected twenty-two subchiefs to represent the many autonomous Native communities that actually occupied this area.[38]

Stevens also designated chiefs to represent groups he identified as the Snoqualmie, Snohomish, Lummi, Skagit, and Swinomish "tribes," who would all be covered by the same treaty as the Duwamish and Suquamish. He used the same process to convene other Native groups to make separate treaties to the south and north. Stevens planned to bring all of the Puget Sound area under American control through three treaties to be signed at Medicine Creek (Nisqually Valley), Point Elliott (Mukilteo), and Point

No Point (northern Kitsap Peninsula)—all within the space of one month. Next, Stevens planned to secure treaties with Native peoples along the coast and east of the mountains. All of these areas had settlers whose land claims had yet to be legitimized under US law.[39]

The terms of the treaties were broadly outlined before the signing conventions: the US government would buy Native land for its American settlers and would provide schools, health care, and business opportunities to the region's Native people. At the first treaty convention at Medicine Creek, however, it quickly became clear to the Native representatives that the devil was in the details. The tribes and the American government, in the person of Governor Stevens, had very different ideas about what constituted a fair exchange.

When the chiefs convened on December 26, 1854, Stevens had already written up the treaty, without consultation, for them to sign. It designated three small areas of land reserved for the nine southern Puget Sound "tribes," including the Nisqually, Puyallup, and Squaxin, within whose territories the reservations would be located. In return for ceding the rest of their lands, the United States offered $35,000, to be paid in kind as goods and services over a period of twenty years. For their part, the region's Native people had to vacate their lands within a year, reduce their combined stock of horses to five hundred, and submit to being governed by the United States. They also had to give up their slaves and refrain from fighting with each other, but they could continue to fish and hunt in all of their "usual and accustomed areas." Stevens made clear that he was not there to negotiate but to issue an ultimatum: take the deal or leave it.[40]

Some of the appointed chiefs chose to leave it. Those who did sign the treaty made a simple mark next to the phonetic representations of their names on the treaty document. One of Stevens's designated chiefs for the South Sound tribes was Leschi, the Nisqually leader. According to many historians, Leschi had worked loyally for the British at the Hudson Bay Company for many years, but he declined to sign the treaty with the Americans. It is believed that Stevens's men forged his mark so that the treaty could be ratified. Whether or not he signed the document, Leschi's bitter sense of betrayal and his anger at the dispossession of his people would cost the Americans dearly in the months and years ahead.[41]

After the signing at Medicine Creek, Stevens moved on to central Puget Sound and the gathering of its tribes at Point Elliott on January 22, 1855. Meanwhile, news of the Treaty of Medicine Creek had spread. Many of the Point Elliott treaty subchiefs and the people they represented balked at the terms that had been forced on the southern tribes. Word also leaked out that the Duwamish people would be required to abandon their homelands and move north or across the Whulge. This demand particularly angered the bands living along the White and Green Rivers, who objected to leaving their ancestral lands. Discussions between Stevens's and Se'alth's representatives had failed to reach any agreement to establish a reservation in traditional Duwamish territory. As the date of the convention neared, tribal leaders arrived to discuss the treaty they were expected to sign. Se'alth himself did not show up until the day of the convention, and it is rumored that he threatened not to attend at all.[42]

The treaty ceded fifty-four thousand acres of Duwamish land to the United States. In the end Se'alth added his signature, but only three of his twenty-two subchiefs signed. This lack of support greatly undermined the authority of Se'alth's reluctant agreement. Most of the upriver Duwamish village groups refused to cede their lands. Stevens, however, was unfazed, considering Se'alth's signature sufficient to validate the treaty.[43]

Three days later, Stevens met the remaining Puget Sound tribes and the Olympic Peninsula's S'Klallum leaders at Point No Point, on the northern tip of the Kitsap Peninsula. Some headmen objected to the amount of land the Americans wanted and to the location Stevens had chosen for their reservation. "I don't want to sign away my right to the land. Take half of it and let us keep the rest," the Skokomish chief Hool-hol-tan demanded. "I am afraid that I shall become destitute and perish for want of food. I don't like the place you have chosen for us to live on. I am not ready to sign the paper." The tribes did not sign the treaty that day, but they reluctantly agreed after discussing their options—or lack thereof—overnight.[44]

Having secured the paper rights to millions of acres, Governor Stevens turned his attention to the other side of the mountains. The treaties he extracted from the eastern tribes in the spring and summer of 1855 would have consequences as least as significant, if not more so, for the Puget Sound settlers, especially those who laid claim to land in the largely unceded

Duwamish watershed. "Seattle and other headmen had invited Americans into their country to live together and share in prosperity," explains David Buerge in his biography of Se'alth. "But the treaty stipulated that upon ratification the people had one year to move away from growing centers of commerce to isolated reservations. As this sank in, anger grew."[45]

The brunt of the resistance to the treaties—led in large part by Leschi of the Nisqually and Kamiakan of the eastern Yakama—would play out almost entirely in the Duwamish watershed. A storm was brewing, and it was aimed straight at the river's settlers who had been so warmly welcomed by Se'alth just a few years before.

After Governor Stevens concluded his eastern Washington treaties, unrest erupted across the mountains. Shared anger and family ties between the Yakama and Leschi's Nisqually tribe formed the basis for an alliance as both groups struggled to retain control of their lands and resources. Many of the Duwamish people, who had not been provided with a reservation in their own homelands, had family ties to the Yakama, and as most of their designated subchiefs had refused to sign the treaty, they felt no obligation to abide by its terms. As Leschi began to coordinate with these and other tribes angered by Stevens's demands, they became more determined to expel, or at least contain, the settlers intruding into their lands.

The eastern tribes had been assured that they would not have to leave their lands until after the treaties were ratified by the US president in two to three years, and that no settlement by whites would be permitted until that time. Within two weeks of the treaty signing, however, Stevens announced in the Oregon and Washington newspapers that the eastern plains were open to settlement. His proclamation set off an immediate land rush, with violent consequences.

The conflicts were partly driven by the discovery of gold along the Colville River. Incursions by prospectors violated the treaty promises made by Governor Stevens. In September 1855, the Yakama killed a group of miners who encroached on their lands east of the mountains. Bigelow's Black River Coal Company partners were probably also killed, as they had been scouting for gold in the area and never returned. The pioneer settler Arthur Denny had been on his way to survey an improved road route over the mountains when he received word of the attacks and turned back. The Indian agent Andrew Bolton, who was sent to quell the hostilities, was killed as well.[46]

In Stevens's absence, acting governor Charles Mason sought Leschi's advice on dealing with his Yakama relatives, but Leschi told him that the Nisqually were equally dissatisfied with the conditions of their own treaty. Leschi demanded reservation land adjacent to the Nisqually River and was denied.[47]

In October, Leschi sought the help of James McAllister, one of Puget Sound's earliest American settlers, whom Leschi had befriended in 1844. Leschi explained that his people could not possibly survive on the meager reservation assigned to them and would fight rather than settle there. Without access to their river and prairies, the Nisqually people were afraid they would starve.[48]

As Leschi's frustration grew, he began to organize a group of resistance fighters. McAllister wrote to the superintendent of Indian affairs in Olympia to warn of the growing threat posed by his former friend. "[Leschi] has been doing all that he could possibly do to unite the Indians of this country to raise against the whites in a hostile manner and has had some join in with him already," McAllister wrote in October 1855. "Sir, I am of the opinion that he should be attended to as soon as convenient for fear that he might do something bad. Let his arrangements be stopped at once."[49]

McAllister offered to lead a team of volunteers to locate and appease Leschi. He set out with instructions to bring the Nisqually leader to Olympia. The stakes were high, as Mason had issued instructions to kill any Indians who resisted capture. Leschi's brother Stoki went with the McAllister party as their guide, but he was loyal to the Native fighters and acted as an informant to prevent his brother's capture. The group set out to search for Leschi, who fled his Nisqually home for the upper Duwamish watershed's White River, where allied fighters from several tribes were gathering. The White River led to Naches Pass, long used by tribes on both sides of the Cascades as a trading and raiding path through the mountains.[50]

A couple of weeks earlier A. L. Porter, a White River settler, aware of the growing tensions, had taken to sleeping in the woods near his cabin to protect himself from a nighttime ambush. On September 27, 1855, Porter was awakened by a group of Native fighters ransacking his cabin. In the morning, he and his neighbors fled to Seattle. Mason was summoned, and after stopping to question a group of Native men near the White River's Muckleshoot Prairie, he insisted that there was no threat to the settlers. Reassured, many of them soon returned home.[51]

The search party led by McAllister was ambushed the next month while tracking Leschi, and McAllister and another settler were killed. Despite their previous friendship, Leschi reportedly ordered the fatal shot that killed McAllister. He no doubt knew that under Mason's orders, the alternative was to risk being captured and hanged.[52]

The following day, Leschi's forces attacked the American settlers who had returned to their homesteads. Mason's assurances notwithstanding, the fighters were determined to drive the Americans off their lands. Several of the attackers had been friendly with the victims, but the looming threat of dispossession and starvation made the Native forcers desperate and bold. Nine people from three settler families were killed and their bodies brutalized in a warning to downriver settlers. All the adults were killed in the attack; three children were helped to safety by a Native man who found them wandering in the woods. It is uncertain whether the attackers had set the children free because they took pity on them or in order to deliver a message, but when the children made it downriver, they sent a panic through the settlers with their horrific tales of the ambush. Years later one of them, John King, who was seven years old at the time, described being "terror-stricken" by the massacre.[53]

Within a month, all "friendly" members of the Puget Sound tribes were ordered to a few tightly controlled internment camps on islands and peninsulas across the water. They were warned that if they refused to go, they would be treated as hostile by the Americans and could be killed. Exceptions were made for workers at Henry Yesler's mill in Seattle: their encampment was in the city, next to the mill, where they could continue to work and keep the mill running. The few trusted Native couriers and others who were permitted to leave the camps were required to carry papers showing they had permission to travel.[54]

Despite the threats, many Duwamish people still refused to leave their homelands, staying in their well-stocked winter villages around Lake Washington. In late December, Se'alth and his Suquamish followers complied with the order, making the rough winter journey across Puget Sound to their assigned camp at Port Madison. Most of the Duwamish stayed behind or agreed to decamp temporarily to nearby Bainbridge Island. They would return, staying at Black River under the leadership of the Duwamish leader Studah (aka William) once the fighting died down. They felt no obligation to abide by a banishment order under a treaty they had not signed.[55]

The fighting threatened the political and economic alliances that Se'alth and his allied chiefs had spent years attempting to build with the settlers. The war was destroying their vision of coexistence between whites and Natives. While Se'alth is said to have considered joining the resistance fighters himself, he eventually threw in his lot with the Americans.[56]

At Port Madison, the Americans built a fortress they called Fort Kitsap, after Se'alth's uncle. Off and on for the next month, snipers crept into the camp and tried to get a shot at Doc Maynard, who had been appointed the Indian agent in charge of Port Madison. Se'alth and his Suquamish allies, always on alert, succeeded in keeping Maynard safe. Then in late January, Se'alth received word that Native forces were gathering on the east shore of Lake Washington: Yakama fighters from both sides of the Cascades were joining Leschi's Nisqually and upper Duwamish-White River fighters. Among them was a Yakama-Duwamish headman named Klakum, and Owhi, a Yakama headman who reportedly had killed Se'alth's nephew in 1841. Ada Smithers, the daughter of a Black River settler, later wrote that during this time Leschi and Owhi were both living at a Native village on her family's farm. The Native troops were commandeering a fleet of canoes from the Duwamish villages around the lake and planned to attack the settlers in Seattle in the coming days.[57]

Se'alth and Maynard paddled across Puget Sound to warn the military commanders of the impending attack, but they returned to Port Madison in frustration when their warnings were dismissed. On the evening of January 25, Eli Maple was visited by a local Duwamish man known as

Salmon Bay Curley, who warned that 2,500 fighters were assembled near Seattle. The following morning, the Native forces attacked. Beyond the clear-cuts marking the edge of the small settlement, the allied fighters fired round after round. They had the advantages of forest cover and elevation on their side, but the Americans had a warship in the bay, cannons, and a seemingly endless supply of ammunition. At the end of the day, bullets spent, the Native forces retreated upriver, burning all of the Duwamish River homesteads as they went.[58]

Se'alth and the Suquamish could hear the battle rage from their camp across the sound at Port Madison. The Duwamish people who had stayed in Seattle to work paddled their canoes out into the bay, far enough off-shore to wait out the battle in safety. Writing about the war years later, Emily Denny lamented, "Cupidity, race prejudice and cruelty caused numberless injuries and indignities against the Indians." But, she wrote, "In spite of all, there were those among them who proved the faithful friends of the white race." On the other side, those Native people who joined the resistance—and even some who didn't—left behind descendants who, several generations later, still considered Se'alth a traitor to his people.[59]

There was no clear victor in the Battle of Seattle. The settlement survived, and only two settlers were killed. "A few of the settlers had the courage to return to the wreck of their homes," wrote Clarence Bagley. "The land was there; the small clearing was there and even the young orchards remained . . . but the houses, barns, fences, and livestock had been swept away. It was beginning life over again." Many of the settlers left when the war was over, including Eli Maple. Perhaps more significant, the war had scared away other would-be settlers and investors.[60]

The conflict lasted only a few months before it was formally ended by an accord reached at the Fox Island encampment near Tacoma. The fighters' demands were partially satisfied: the Nisqually and Puyallup and their upper Duwamish allies from the White and Green Rivers gained new and expanded reservation lands along the White, Puyallup, and Nisqually Rivers. In 1929, Bagley wrote, "The terms granted were far better and more satisfactory to the Indians than what were offered prior to the breaking out of hostilities."[61]

Reviewing the events more than a century later, the American Friends Service Committee agreed, noting that "the changes belatedly set right most of the grievances which had, among other things, caused the war in the first place." The status of the lower Duwamish people, however—those who allied themselves with the settlers—remained unchanged. The historian Archie Binns summarized their fate in 1941, writing that the lower Duwamish and other "Indians who stayed at peace suffered the most." They were not given a reservation in their homeland, and were banished from the Duwamish Valley to distant reservations with other tribes.[62]

In the decades following the war, ninety-plus Duwamish longhouses lining the watershed's rivers and lakes were razed and burned. The Duwamish village of Herring's House, near the Denny Party's original West Seattle landing, was the last to go, burned to the ground in 1893. The *Seattle Press-Times* reported that those who had lived there were arriving at an encampment called Ballast Island on the Seattle waterfront. They blamed the arson on a white man named Watson and lamented the effect of the fire on the village's elders. In an understated condemnation, the paper opined that the displacement of ill and elderly residents "makes the affair look inhuman."[63]

ASSIMILATE OR SEGREGATE

By the time of the Maple-Cavanaugh family's winter feast in 1866, a decade of relative peace, or at least resignation, had elapsed since the war. Most skirmishes between the tribes and the settlers west of the Cascades had ceased, and the majority of the region's Native people had moved to the reservations. In a stinging rejection of Se'alth's vision of coexistence, Seattle's first official act when it incorporated in 1865 was to prohibit Native people from living in the city, unless as laborers housed by their employers: "Be it ordained . . . that no Indian or Indians shall be permitted to reside, or locate their residences on any street, highway, lane, or alley or any vacant lot in the town of Seattle." Assigned to reservations outside their homeland and banished from the city they had helped to found, the Duwamish had few options: merge with another tribe, assimilate into settler society, or go underground.[64]

Some Duwamish, particularly from the upper watershed, moved to the new Muckleshoot Reservation on the White River, which they shared with members of the upper Puyallup Tribe. Others moved to reservations where they had kinship ties, including Se'alth's Port Madison Reservation across Puget Sound, the Tulalip Reservation near Everett, and the northern Lummi Reservation near Bellingham. Many Duwamish people, however, still chose not to move. A large number of Duwamish people returned to their Black River homeland under the leadership of subchief William, who had refused to sign the treaty. Unwilling to leave the land where their ancestors were buried, they either lived and worked in and near the towns or retreated farther upriver to avoid conflicts with the settlers. Some continued to hold out for a reservation of their own, while others pursued Se'alth's vision of coexistence with the Americans.

As early as 1857, the Indian agent Thomas Paine wrote to the Indian Affairs superintendent that tensions between the Suquamish and Duwamish at Port Madison made it difficult for the two groups to share a reservation: "I would therefore suggest, most respectfully, that the D'Wamish Indians be allowed a reservation on or near the lake fork of the D'Wamish River. This tract of land has been cultivated for years by them." He concluded, "Considerable anxiety is felt by these Indians on account of the failure of the Government to confirm to their treaties." Nothing came of his request.[65]

Many of the Duwamish who remained on the Black River lived in a village at the junction of the Black and Cedar Rivers, on the land claim of the settler Erasmus Smithers. Members of the tribe worked for Smithers and other local farmers, but some settlers complained that the Native residents should be removed to their assigned reservation. When US Indian agents responded, Smithers defended the Duwamish, insisting that they were living on his land with his approval and arguing that they provided security for his wife and children when he was away.[66]

In a last-ditch effort a full decade after the Indian Wars, the remaining Black River Duwamish, numbering about 275, once again lobbied the superintendent to put aside a 160-acre reservation for their use—a piece of land no larger than the property claims granted to single male settlers. In 1865, the newly appointed Washington superintendent W. H. Waterman visited the Black River settlement. Convinced of the rightness of their cause, he appealed to the US commissioner of Indian affairs to grant their request. Writing in September 1865, Waterman cautioned, "The Black River Indians, residing near the confluence of the Black and White rivers, claim that they were not represented in the treaty of Point Elliott. They are unwilling to leave their present place of abode, and ask to have a small tract of land there reserved to them. . . . My own judgment, after visiting them and counseling freely with them, is that nothing less than the reserving of the land where they are, and the guaranteeing to them the right to remain on it, will satisfy them."[67]

In a surprising attempt to deliver justice to the disenfranchised Black River Duwamish, Waterman pressed on. "The white settlers in the neighborhood desire to have them remain among them, that they may avail themselves of their labor, yet at the same time they are unwilling they should have a reservation where they are, because they, the white men, want to appropriate the valuable bottom land which they occupy." Over the settlers' objections, Waterman's recommendation to the commissioner was clear. "I have no doubt of the propriety of giving the Indians a small reservation at that place."[68]

King County's settlers squelched the proposal. They responded with a petition objecting to the reservation as "injurious" to themselves and "unnecessary to the aborigines." The settlers argued that the Duwamish people's "interests and wants have always been justly and kindly protected by the settlers of the Black River country." Nearly all adult white men in King County signed the petition. Erasmus Smithers was not among them, although the establishment of a reservation might have caused the forfeiture of his own land. His abstention was moot: faced with nearly unanimous opposition from the county's settlers, the superintendent's proposal for a Duwamish reservation on the Black River was rejected.[69]

With that door closed, the Duwamish families remaining on the river had to seek new ways to stay in their homeland as the number of immigrants once again began to climb. In 2015—150 years later—the Duwamish chairwoman, Cecile Hansen (formerly Cecile Maxwell), would reflect, "We sacrificed our land to make the city of Seattle a beautiful reality. We are still waiting for our justice." Ken Tollefson, an archaeologist assisting the tribe, succinctly described the plight of the remaining Duwamish, calling them "refugees in their own land."[70]

In addition to Se'alth's family, the descendants of Tupt-Aleut and her husband Kruss Kanum were among those who chose the path of coexistence. They married American settlers and learned to navigate the US legal system, and some ultimately gained title to lands in traditional Duwamish territory. Tax records, deeds, and probate documents provide evidence of how these Indigenous families navigated the transition from traditional communal-resource users to private property owners. As a result, 170 years after Tupt-Aleut and Kruss Kanum joined Se'alth in welcoming the first settlers, their mixed-race descendants still own the house in which their daughter, Quio-litza, drew her last breath. Quio-litza's great-great-grandson, James Rasmussen, lives there today. "There is a reason [we] have settler ancestors," Rasmussen told the Native ethnographer Julia Allain in 2014. "Without that, we couldn't have stayed here."[71]

Quio-litza's marriage to R. M. Bigelow lasted only eight years. He died in 1861 but was remembered by local Duwamish families. Some of them adopted his name, including Quio-litza's uncle Sdabahld, a shaman and traditional doctor who became known as Dr. Jack Bigelow.[72]

Quio-litza's second husband, Abner Tuttle, was also an American settler. He seems to have been a man of lesser means and reputation than Bigelow. He first appears in the census records in 1860 at Port Madison, the site of the Suquamish Tribe's Old Man House. Tuttle worked as a lumberman and engineer at the mill at Port Madison and may have met Quio-litza and her father there. After marrying in 1862, Quio-litza and her husband lived at Dogfish Bay, near Port Madison, and then later at Salmon Bay, north of Seattle. When their first child, Nellie, was six years old, the couple moved to Vashon Island, a rural island southwest of Seattle, where Abner Tuttle planted the island's first apple orchard. The orchard still stands today on Backbay Creek, next to Quartermaster Harbor. Nellie was followed by three sisters and two brothers.[73]

Vashon Island census records show that their biracial family was far from unique. The 1871 census documents many mixed Native and white families living nearby, as well as an impressive diversity of immigrants from

Denmark, Russia, Canada, Norway, China, England, and the American states. The new arrivals homesteaded, farmed, and worked as lumbermen, traders, schoolteachers, cooks, and—like Tuttle—fishermen and farmhands. The first generation of Native-settler children on Vashon Island was raised in a multiethnic and multicultural environment that is difficult to imagine there today.[74]

Many of the early settlers of Puget Sound were not wealthy: they had come to escape hardship and poverty in the Midwest. The American government's offer of free land attracted settlers from across the United States, including many who had hoped (most of them in vain) to strike it rich in the California gold rush. Finding themselves out west and cash poor, a steady stream were attracted by the promise of free land up north. As the memory of the Indian Wars faded, more and more immigrants arrived in the Northwest.

As the eldest of the Kanum-Tuttle children, Nellie was the first to leave home. At age fifteen, she returned to the land of her Duwamish relatives. By then, nearly 3,500 white settlers lived in Seattle. Dozens more settler townships lined the valleys of the Duwamish, White, Green, Cedar, and Black Rivers.[75]

It was common practice during this time for older children and young adults to work in wealthier families' homes in exchange for room and board and the opportunity to attend school or earn a small wage. Most of the region's wealthier immigrants settled in Seattle and pursued fortunes in exports, land speculation, and railroad construction, or as doctors, lawyers, and engineers. No documents survive to show where Quio-litza and Bigelow's children lived during her years on Vashon Island, but as the Tuttle children began to leave home, their half-siblings reappeared in the public record. In 1879, Nellie and her half-brother Miles Bigelow Jr. were serving together in the home of a recently arrived young attorney and future railroad magnate, the newly appointed Judge Thomas Burke.[76]

Nellie and Miles soon relocated to work as a house servant and farmhand in the small town of Lake Washington. They were now just a few miles from their grandmother's ancestral village of Sbabadid. Despite Nellie's brief term of service in Burke's home, the historical record suggests that she learned a few things about the settlers' land laws from her employer.

Dr. Jack Bigelow, a Duwamish healer. Bigelow was Quio-litza's uncle; he adopted his westernized name from her first husband, Dr. Ruben Bigelow—a medical doctor and coal entrepreneur who settled in Renton in the 1850s. Courtesy of the Museum of History and Industry, Seattle.

Their paths would cross again in future years, though whether by coincidence or design remains unknown.[77]

While Quio-litza was living on Vashon Island, her sister Ellen married an American settler who bought land at Black River Junction—the confluence of the White and Black Rivers. Her husband, Gerald Proctor, owned several properties in and around Seattle, including at Ross (now Fremont) and Green Lake. But it was the land at Black River Junction that allowed Ellen to remain close to her family, including, perhaps, to Quio-litza's children from her first marriage. It appears that Ellen did not leave the Black River area and was still living there when Nellie returned with Miles Jr.[78]

Nellie's great-uncle, "Dr. Jack" Bigelow (Sdabahld), was also there to welcome her home. Dr. Jack stayed close to the Black River, living out his life in the Cedar River Valley above Lake Fork. As a prominent shaman, he occasionally appeared in news stories and historical records. He

Quio-litza (Ann Kanum) petting a deer on the porch of the
house she shared with her second husband, Abner Tuttle.
Courtesy of the Rasmussen/Nelson family.

was also featured in ethnographic reports shortly after the turn of the
century, as anthropologists recorded the local Indigenous traditions that
were fast disappearing.[79]

Dr. Jack was the only native person to secure a land claim from the
government in traditional Duwamish territory. In 1884, the United States
supplemented its Homestead Act to allow claims by Native people who
agreed to renounce their tribal affiliation and become naturalized US citizens. In 1891, perhaps with Nellie's help, Dr. Jack used this law to apply
for a claim of his own on the Cedar River. He was ultimately granted fifty
acres. Far from renouncing his Native identity, he erected a longhouse on
his land and hosted ceremonial gatherings there until the end of his life.[80]

In 1896, Dr. Jack hosted the last recorded bone game—a traditional
gambling contest (also known as a sing gamble) practiced long before
colonial settlement. It was attended by a reporter from a local newspaper,
who observed, "None but the oldest Indians took part in the game, for
only the patriarchs of the tribes knew anything about it, but the squaws

and young bucks were permitted to assist in the preparations, and when the game commenced, joined the choir and aided in the 'music.'"[81]

The reporter described a high-stakes gambling game between the Duwamish of the Black and Cedar Rivers and their Puyallup adversaries, who had traveled north for the occasion. Two hundred Puyallups arrived in sixteen wagons, as well as in buggies, on horseback, and on foot. They competed for horses, wagons, saddles, blankets, jewelry, rifles, clothing and cash. "In the event the Puyallups lost they would have been compelled to walk home," the reporter wrote. "But if the Black and Cedar rivers had lost, it wouldn't have done them any good to walk, because they would not have any home to go to."[82]

At 8 p.m. on a Monday, the game began. Inside his "teepee," as the reporter called it, Dr. Jack shuffled eight white cedar chips and one black one. The competitor for the Puyallup correctly tapped the ball in which Dr. Jack had hidden the black chip, earning one of the sixty-six points needed to win the game. A correct choice earned points, but a wrong choice lost points, making for some very long contests. Twelve hours after the competition had begun, the Puyallup had scored fifty points, but at the end of twenty-four hours, they were almost back to zero. After three days of nonstop play, Dr. Jack relinquished his position to another dealer, and the game continued. At the end of five sleepless days and nights, the two sides called the sing gamble a draw.[83]

Dr. Jack died on July 4, 1901. Most of his land was passed on to Quio-litza and her sister. Quio-litza's daughter Nellie and her husband, Myron Overacker, paid the taxes. Dr. Jack's widow sold a small section of the flats by the river to a coal speculator, who abandoned the property when he found that the coal vein he coveted was in the hillside owned by the Duwamish sisters. It is not clear whether the coal on the sisters' land was ever worth mining, but the Pacific Coast Coal Company operated the New Black Diamond Mine—also known as the Indian Mine—across the river from their land between 1927 and 1941.[84]

By the time Dr. Jack passed away, the rest of the family had been reunited in Duwamish territory. Tupt-Aleut, Quio-litza, and Ellen all lived on the east side of Lake Washington, and Nellie served as an informal property manager and legal adviser to her extended family.[85]

On a high knoll across the lake, Nellie and her husband were busy creating their own homestead. Myron Overacker secured a piece of land atop Beacon Hill, just south of the Seattle city limit. The property had been allocated to the University of Washington, but it was considered surplus and sold after the state built its new campus in north Seattle. This became the couple's home for the rest of their lives. Three generations later, Nellie's granddaughter, Ann Rasmussen, raised her own children in the home, steeped in the stories of their ancestors and of the homestead that sheltered their family in the years after the treaties dispossessed the Duwamish of their lands.[86]

Sweat and the Transformation of a Watershed

Moving Mountains and Rivers

1900–1950

AS THE KANUM-TUTTLE FAMILY WERE RESETTLING ON THEIR ancestral lands around Lake Washington, a new immigrant family settled on land straddling a stream known simply as "the outlet"—a small tributary connecting Lake Union to the tidal inlet called Shilshole by the local people, and Salmon Bay by the white settlers.[1] John Ross claimed the property, north of Seattle, before the Indian Wars. When fighting broke out, a local Duwamish man warned Ross of an imminent attack on his homestead, and Ross fled to the safety of Seattle and its bunker. He was close enough to see his house go up in flames as he fled. He intended to return and farm the land he had claimed, but the unrest led him to spend several years away. He served in the settler's militia while the war raged and then retreated to Oregon. While there, he met and married Mary Jane McMillan. The couple returned to Seattle to start a family. Ross worked in Henry Yesler's sawmill, working to raise the money he needed to rebuild his house.[2]

The Duwamish Waterway's straightened canal shortly after its construction, with some original riverbends still unfilled (ca. 1922). In the following years, the meanders and surrounding lands were developed into Seattle's industrial district. Courtesy of the *Seattle Times*.

When Ross's daughter, Ida, was twelve, the family finally moved onto his land claim. Ross planned to build the new family home by hand and to produce their own food. Ida had warm memories of sitting by the fire in their two-room log cabin, helping her father make bullets for his rifle. "We would all gather around Father on the floor and watch him let the lead . . . and pour it from a little long-handled iron spoonlike thing into bullet molds," she wrote.[3] Among the other things Ida recalled fondly was "the most delightful, beautiful little stream winding in and out among the trees and overhanging brush and filled with all kinds of little fish and bugs and frogs that sang such a lovely little song evenings in the spring." The 160-acre property straddled the narrow outlet stream. "It was a paradise to us children," Ida wrote. "We were very happy in our new home. Then came trouble for Father."[4]

In 1883, a group of investors in the Lake Washington Improvement Company petitioned for a right-of-way through John Ross's land claim in order to cut a canal for transporting logs between the freshwater lakes to the east and the saltwater inlet to the west. Their canal would extend from Lake Washington through Lake Union and on to Salmon Bay via the creek on Ross's property. Arguing that the canal was in the public interest, they asked the local justice of the peace to grant them the use of one hundred feet on either side of the creek to dig their canal.

Ross fought back. According to Ida, "He wasn't a progressive man, didn't want to be disturbed, and naturally, didn't want the creek to be disturbed either." Ross asked the court for an injunction to stop the company from widening his creek. He countered that the plan would harm his property and his ability to farm. If the project were not stopped, he asked that the company be ordered to put up a bond against any damages. But the court determined that any potential damage to Ross's land was outweighed by the benefits of the canal. That Ross did not see the canal as an "improvement" to his property was immaterial.[5]

Defeated in court, Ross sought to stop the canal by more direct means. "The logs coming through would destroy a bridge Father had built," recalled his daughter. When the canal company's workers arrived, "Father thought, very foolishly, that he could stop this work and went down many times with his long rifle and drove them off."[6] The company applied for a restraining order against Ross, claiming that he "threatens to kill the surveyors, engineers, and servants of the plaintiff herein," and that the company "believes that the defendant will carry out his said threats [and] kill some of its employees."[7] In March 1884, the court issued the order, and the outlet creek was widened. The story of Ross's split homestead is just one of many tales of conflict between settlers with competing visions and interests in the waterways that crisscrossed the growing new metropolis.[8]

DREAMING A WATERWAY

Settlers and local business interests shared a desire to reshape the region's waterways, though their objectives differed. Farmers in the Duwamish watershed, frustrated by the river's annual floods, sought relief in the form of impoundments and diversions from the nascent governments and on occasion took matters into their own hands. Using dynamite and makeshift levees, local farmers tried, and mostly failed, to divert floodwaters away from their crops.

Downriver the city's industrial boosters were seeking quick routes into the deep freshwater berths offered by Lake Washington. Investors were also jockeying to purchase property along prospective inland canal routes, anticipating that their value would skyrocket once a canal was built to link Puget Sound to the inland lake. Judge Thomas Burke was one of these, having invested a considerable amount of money in land and a future railroad line running northeast of the city center, which he imagined would follow a deep ship canal. John Ross's land lay along this route. The small stream crossing Ross's property was the basis of the imagined link between the lakes and Puget Sound.[9]

As early as 1854, the pioneer settler Thomas Mercer had envisioned a link from the saltwater inlet of Salmon Bay to the huge lake east of Seattle—then known as Lake Duwamish—and suggested the name Lake Union for the smaller lake in the middle, where a passage connecting the two would meet.[10] The first prospector to own land along the proposed canal route was Harvey Pike, the son of John Pike, a founder of the original University of Washington in downtown Seattle. The younger Pike was hired to clear land that the university owned north of Seattle—one of several parcels Congress had set aside around the city for a future public university campus. Pike was paid in kind for his work clearing the land; he was granted 162 acres of property on the isthmus between Lake Duwamish and Lake Union. He filed a street plat for a new Union City there in 1869.[11]

Pike reserved a two-hundred-foot right-of-way for the prospective canal and is rumored to have begun digging it out himself by hand. In 1871, he sold this ribbon of land to the Lake Washington Canal Company, of which he was a founder. These were the first efforts to connect the two lakes—an enterprise that would eventually transform the entire Duwamish watershed.[12]

Pike's canal scheme faltered in the end, and a later government plan to cut a straight channel directly from Elliott Bay to Lake Union was also abandoned as unrealistic. But a decade later, the persistent Thomas Burke, supported by a small group of investors, again took up the dream. They formed the Lake Washington Improvement Company, surveyed possible routes, and hired contractors to begin digging a channel, including the section through John Ross's land claim. The Seattle historians Jennifer Ott and David Williams chronicled the construction of the resulting canal, following its fits and starts, lawsuits, labor disputes, and political opposition that plagued the project for the next four decades. "In particular, the state Democratic Party opposed what they called the 'Seattle Ditch,'" they wrote. Despite support from the Republican governor, Congress refused to provide funding in the 1892 Rivers and Harbors Act. Without government support, the project once again languished.[13]

The boosters of the canal through Lake Union were not the only ones dreaming of a path to connect Lake Washington to Puget Sound. They had competition. In April 1897, the former Democratic governor Eugene Semple sent an enthusiastic letter to his eldest daughter, Adria, about his life and business in Seattle: "My Dear Daughter. We are now keeping house and have a little dwelling at the Tip! Tip! top of a high! high! hill," he wrote, describing his daily commute by streetcar and the steep three-block walk to the Queen Anne house he had recently moved into. It had grand views of Puget Sound, Lakes Union and Washington, and the Olympic and Cascade mountain ranges beyond. "We can look down on the bay and see our dredgers at work," he boasted, "and down on the City and see the streetcars running in different directions, and at night it is a proud sight to watch the electric lights appear."[14]

Semple's pride in these sights reflected his own sense of importance in the modernization of Seattle. The dredgers he could see from the house belonged to his Seattle and Lake Washington Waterway Company (SLWWC). Two years earlier, with his middle daughter, Zoe, at his side, he had inaugurated his grand plan to cut a series of canals into the tide flats at the mouth of the Duwamish River, and another through the steep hill directly east of them. His goal was to connect Puget Sound to Lake Washington by the shortest possible route. In the process, he would fill the surrounding "foul" tidelands at the mouth of the Duwamish River, which he decried as "valueless for all purposes of navigation . . . and generally unhealthy, unsightly, and worthless for any and all purposes." In their place, Semple would create valuable new land for railroads and industry. He saw himself as helping to usher Seattle into the twentieth century, securing his own financial future in the process.[15]

Thousands crowded the shores as Zoe, then twenty-two, climbed atop the industrial-scale dredger and started its engine. It cut into the mud, pumping saturated fill into an enclosure delineating the lands that Semple would create.[16]

A former newspaper editor, Semple had been appointed governor of Washington Territory by President Grover Cleveland in 1887. He served only two years before being replaced by Cleveland's successor. Semple then ran for election as governor of the new State of Washington soon after but lost to the Republican candidate, Elisha Ferry, and turned his attention instead to a variety of business interests. His time in politics had earned him some influence, as well as his share of opponents, both of which would affect the success of his plans to build what was referred to as the South Canal.[17]

Semple envisioned his ambitious proposal as an alternative to the flagging plan to construct a government-funded canal through Lake Union, which was supported by Thomas Burke and the city's Republicans. In Semple's view, "At the time that the SLWWC appeared on the scene all hope of government aid had been abandoned." Instead, his company proposed to build a privately funded canal and locks just south of the city limits, involving a deep cut through the high point of Beacon Hill and a set of canals draining the tide flats at the mouth of the Duwamish River. Semple argued that he could construct the canal in a few years and fund it entirely through profits from selling the reclaimed land he would create in the tide flats below, using fill from the digging of the canal.[18]

The proposal to redistribute dirt from the hill onto the tidal lands at the mouth of the Duwamish River was considered audacious, but Semple's enthusiasm was contagious. In 1893, a year after Congress rejected funding for the north canal, he secured support from political allies in Washington's newly formed state legislature. At his urging, they passed a bill allowing private contractors to build waterways through state-owned lands.[19]

Semple incorporated his company in 1894 with his former political rival Elisha Ferry as president, perhaps to deflect the animus of investors who might shy away from a project championed by a Democrat in Republican Seattle. Semple then began to arrange financing—in some creative ways—for the work of dredging out the tide-flat canals. He raised over half a million dollars in pledges from Seattle residents, banks, and businessmen who, frustrated with lack of progress on the so-called Government Canal, saw the project as a great public work worthy of popular support. He also tapped family ties in the Midwest to secure investment from the Mississippi Valley Trust Company. Most of the financing, however, was to come from the sale of liens on the future tidelands. This strategy became a temporary liability when the new law authorizing the liens was challenged in court.[20]

Despite popular support, not everyone believed the Beacon Hill portion of the project could—or should—be built. Thomas Burke and other backers of the north canal had extensive financial interests tied up in that route and saw Semple's scheme as a threat. Noting that Burke owned property along the northern route, one of his biographers wrote, "Schemes for a canal running directly into Elliott Bay from Lake Washington had no appeal for him." Burke's father-in-law, Judge John McGilvra, wrote a letter to the local paper lambasting the law that let Semple build his waterway through state land, "appropriating school lands to private use without compensation." He warned that people who were impatient for the north canal had now "become reckless, and are ready [to] engage in any wild scheme that may be presented by political demagogues or financial speculators."[21]

Burke himself is known to have engaged in some creative (and ethically questionable) strategies to clear the titles needed to build the north canal. John Ross's resistance to canal construction through his land had made him a persistent thorn in Burke's side. In 1885, Burke represented Ross's wife, Mary Jane, in her petition for divorce and helped her gain control of the north side of the property in the divorce settlement. John Ross died just a few months later, after writing his ex-wife out of his will. But Burke and his partner, Daniel Gilman, were able to buy the right-of-way for the railway from Mary Jane. Despite the federal government's refusal to finance the canal, Burke and his allies were determined to forge ahead.[22]

Finances aside, Semple's critics questioned the engineering feasibility of cutting a three-hundred-foot-deep canal through Beacon Hill. Some also objected to his plan to charge tolls to ships using the canal, privatizing access to the freshwater lakes beyond. Boosters of both canals wanted access to Lake Washington as a deep freshwater port for use by the US Navy and private industry. The choice of route would determine the fortunes of a great number of landowners, contractors, and investors who had aligned themselves with either Burke or Semple.[23]

The first lawsuit seeking to kill Semple's South Canal was filed in 1894. A year later, the Washington Supreme Court upheld the legality of Semple's state contracts. On behalf of the Great Northern Railroad, Thomas Burke then filed a constitutional challenge to the state law, winning a temporary injunction—until the State Supreme Court once again ruled in Semple's favor. Semple's letter to his daughter captures his optimism at having overcome these hurdles and his progress on filling the tidelands. He included a hand-drawn map of the lands he had already created along the waterfront, including a new island to house the Centennial Flour Mill—a major new industry slated to open in Seattle the following year.[24]

Semple's opponents continued to challenge elements of his plan, and when all legal maneuvers were exhausted, they turned their attention to the project's investors. In 1899, the railroad financier Henry Villard (whose company was a rival of the Great Northern) assumed the presidency of the Waterway Company and invested his own funds. He had difficulty attracting other investors, however, finding "mysterious influences in the channels of reference at Seattle," according to Semple's account.[25]

As Semple met with increasing difficulty in securing the funds for his project, he railed against the "money kings" of Seattle who opposed him. Eventually, a Philadelphia bank purchased $3 million worth of state lien certificates held by the company, restoring his cash flow. The state extended his contracts (which had also been unsuccessfully challenged in court), and the Waterway Company resumed its work, finally tackling the high ridge between the Duwamish tidelands and Lake Washington in 1901. Semple had a contract for city water to power an enormous hydraulic cannon, which he planned to use to sluice through Beacon Hill, sending fill dirt into the tide flats below.[26]

As the canal advanced from the tidelands toward Beacon Hill, Semple's opponents pulled out all the stops. At the turn of the century, Semple found himself and his company at the center of a political, legal and financial firestorm that spread all the way to the halls of Congress in Washington, DC.

At the start of the new century, the north canal backers began to explore local sources of public and private funds to move their project forward without federal money. A small lock and a log flume were installed between Lakes Union and Washington, which enabled the transport of small boats and timber between the two lakes. The King County Council then passed a bond measure of $250,000 for the canal.[27]

During this time, however, ships were growing larger, requiring bigger waterways and causing the cost of the north canal route and its locks between the lakes and Puget Sound to balloon. Semple used this development to advance his own cause, boasting that he could complete his canal and sell it to the government for $2 million—one-third of what it would cost to build the north canal.[28]

In 1902, Congress took the bait. Both Burke and Semple were summoned to Washington, DC, to testify before Congress on the value of their competing canals. Burke fraudulently testified that the Washington State Supreme Court had struck down the 1893 law authorizing Semple's canal scheme. Congress decided to invest a token amount in the construction of a ten-foot-deep log canal between Lake Union and Salmon Bay, following the widened creek through the original Ross land claim. Congress also sent a committee of engineers to Seattle to evaluate the feasibility of both canals. This was the third study of the north canal within four decades, and the first government study of the South Canal.[29]

In addition to being judged on its own merits, the north canal proposal now had a directly competing rival project to contend with. The backers of both projects attempted to promote them by a new measure—plans to relieve flooding along the Black and Duwamish Rivers, where spring rains continued to blow out levees and scour newly settled towns and farms. This addition would prove transformative in the years ahead.[30]

Semple contended that his tide-flat canals would drain the Duwamish River more quickly during heavy rains, alleviating upriver flooding. The north canal proponents altered their plan to lower Lake Washington to the level of Lake Union—a difference of nearly nine feet. The result would save money by eliminating the need for a lock between the two lakes and would stop Lake Washington from flowing into the Black River by lowering the lake level to below the river outlet. Instead, the inland lakes and the water from their 490-square-mile drainage would flow out through the new canal. The change would cleave the watershed in two and remove a vast amount of water that contributed to flooding on the Black and Duwamish Rivers.[31]

In 1903, the Corps of Engineers delivered its report to Congress: it recommended that neither project be pursued. The South Canal's cut through Beacon Hill, while technically feasible, was considered utterly impracticable. The north canal route was judged as far too costly; once again, the government did not believe that it delivered enough benefit to justify federal funding.[32]

Without public funding, the north canal could not be built. Semple's plan, however, was privately financed, so he took advantage of Congress's decision to proceed without competition. He continued to power-wash Beacon Hill into the Duwamish tide flats, but the large volumes of water required to dislodge the hard clay of Beacon Hill caused objections from ratepayers who were effectively subsidizing the company's low water rates. Then a neighboring property owner filed suit against the Waterway Company in 1904, when the cut threatened to swallow his home. Semple won this lawsuit, as he had all the others, but not before the growing political firestorm (likely fanned by Burke) led the City of Seattle to cancel the water contract needed to run the company's dredge. Without water, the canal could not be cut, and Semple's project was dead.[33]

In addition to the lawsuits, Burke had piled up additional barriers to Semple's ambitions, including convincing the railroads that crossed the flats to fill their own tidelands instead of buying reclaimed land from the Waterway Company. A curious confluence of events suggests he may have had additional maneuvers in mind if the canal had not been stopped when it was. In 1897, just as Semple's work on the tide flats was getting underway, Myron Overacker moved to Seattle from Tacoma. He and his fiancée, Nellie Tuttle (Quio-litza's daughter), who had served in Thomas Burke's home years before, settled on state-owned land at the top of Beacon Hill. The land they homesteaded (the state-owned university reserve land that

McGilvra had complained was being appropriated for private use) was right in the path of Semple's planned canal.[34]

If Burke didn't actually arrange for the couple to purchase the land in the South Canal's right-of-way, he may at least have put in a good word with his fellow members of the university's board of regents. Or perhaps the state had simply abandoned its belief that the canal would be built through Beacon Hill after the Army Corps of Engineers declared it was impracticable. But the fact that Nellie had lived in the home of Thomas Burke—the South Canal's most fervent opponent—and wound up living in the path of his rival's canal seems an extraordinary coincidence.

In 1903, the year of the Corps's report to Congress, Washington State signed the contract to sell its university land to the Overackers. After the South Canal project was canceled, the Overackers paid off the loan on their land four years early, receiving the owner's deed to the property in 1907.[35] A portion of their estate remains in the lands of Nellie's Duwamish descendants to this day.[36]

Despite the failure of his canal, there is no question that Semple realized a great many of his ambitions to raise Seattle to national prominence. While making his case for the South Canal to the Army Corps of Engineers, Semple wrote of the tide flats, "On this land will be founded the major portion of the manufacturing industries of the city, and these will be of great benefit and will play an important part in making this city the metropolis of the Pacific Coast." This ambition, he claimed, was driven by a desire for the common good: "The Waterway Company feels that it, too, is benefiting mankind by making solid and substantial land where formerly there was water, the rising and falling of the tides, and occasionally bare and unsightly mudflats, which were a menace to the health of the dwellers on the adjacent dry lands."[37]

Ironically, as governor, Semple had been one of the first to sound the alarm over diminishing salmon runs passing through those same tidelands. Not knowing much about salmon biology, Semple attributed the problem wholly to overfishing. Having tried—and failed—to secure support for fishing restrictions, he was unaware that his later endeavor to fill the Duwamish "mudflats" possibly did more to degrade local salmon runs than any fishing pressures existing at that time. To Semple, taming the region's waterways represented only progress.[38]

Some of Semple's opponents believed that he never really intended to build the South Canal but had proposed the project only to make money by filling and selling liens on the tidelands. Had that been his intention, however, it seems unlikely that he would have lobbied for a bill requiring waterway construction or sought funding that was collectable only upon "the first ship sailing through the canal." Semple himself refuted this interpretation of his goals in his case to Congress, arguing that the company's name and activities always reflected its true purpose. In 1902 he wrote, "Concerning the intention of the company to excavate a ship canal through the hills to the lake, . . . those who desire visual evidence of the true intention of the company should visit the canal right-of-way and watch the actual progress that is being made by way of clutching out the canal through the first hill." Indeed, when the canal was finally defeated, he resigned his post as company president. "The fire of the opposition has always been concentrated against me, and I have fought them back, with success, so far as the issue of the battles in the Courts, the Legislatures, and in Congress have been concerned, but in the financial field I have been, at last, worsted."[39]

Semple lamented the reduced aspirations of the Waterway Company, writing in 1905 that they would now "'toil like Calahan' in the common place business of filling the tide flats." He added that he hoped at least that the company could finally make some money, but that was not to be. Semple died of pneumonia in 1908, leaving his shares in the company to his children. His probate papers state that "the work of said company in filling in the tide flats has been greatly hampered by litigation, [and] the holdings of the deceased in said Company have become utterly valueless."[40]

Semple's most substantial legacy is the filling of the tidelands at the mouth of the Duwamish River and the creation of the industrial district that fueled the growth of Seattle. At the time of his death, his company had filled at least 544 acres of tidelands, and by 1917, when its contract was canceled, it had filled 1,400 acres, creating Seattle's future industrial lands and changing the mouth of the Duwamish River forever. Given Semple's singular role in achieving these ends, it is curious that his name is not more prominently memorialized in Seattle today. Just as the influence of Collins, Maple, and Van Asselt as the city's first settlers is often neglected, Semple's role in transforming the Duwamish River into Seattle's industrial powerhouse is commonly overlooked in our "founding stories."[41]

MOVING MOUNTAINS AND RIVERS

In late November 1906, William Hempel was patrolling the bank of the White River, rifle at his side as a defense against angry farmers from the south. An itinerant day laborer and miner, this week Hempel was a hired gun—one of several who had been sent by the King County sheriff to keep watch over the river. Just a week earlier, the flooding river had spanned a width of three miles. In the last few days, between the town of Slaughter (now Auburn) and the Muckleshoot Indian Reservation, the river had run dry. Its waters had turned south into the small Stuck River channel and the Puyallup basin below. The volume of water overwhelmed the narrow Stuck Valley, sweeping away homes, roads, and bridges. The diversion drained the White River floodwaters, revealing a massive logjam that had forced the river to jump its banks and carve a new channel. King County sent a squad of armed guards to ensure that the Pierce County farmers now in the path of the flood wouldn't dynamite the riverbank to force the water back. The new course of the White River suited King County and its farmers just fine.[42]

The tide-flat canals that Semple constructed were meant to quickly drain the floodwaters coming down the Duwamish River and provide relief to upriver farmers. But the ecology of the watershed had always been reliant on seasonal flooding to replenish and nourish the broad plains where the settlers planted their crops. The settlers saw this low-lying area as ideal for settlement. They cleared nearby forests to create open fields for their farms, ignorant or uncaring of the forests' role in buffering high river flows and absorbing floodwaters. The patterns of erosion and deposition that created the fertile plains were disrupted. Without forest cover, the plains were scoured and their rich soil washed away.

Semple's drainage canals through the tide flats at the river delta were wholly inadequate to alleviate the vast watershed's natural flooding, even without the changes wrought by the settlers. As new immigrants concentrated along the transportation corridors provided by the Duwamish, Black, White, and Green Rivers, flooding intensified. One of the worst floods prior to 1906 had come in the fall of 1892. "The entire Duwamish valley is a great lake," the local paper had reported, colorfully adding that the flood was "the worst in the memory of man—red or white—living on [Puget] Sound." That flood was followed six months later by "a plague of water," threatening economic destruction. "It is thought that many farmers on the Duwamish will lose their entire crop of hay and hops."[43]

The Duwamish Valley floods intensified as more land was cleared. The King County engineer, O. F. Wagener, warned that floodwaters would cascade into the rivers more quickly each year, "in the same ratio in which the land is cleared and thus made to shed its surface water in as many hours as it did take days when the land, in its uncultivated state, was covered with brush, moss, leaves, and timber."[44]

The 1906 flood on the White River claimed the lives of two men trying to ford the river on horseback. The rising water washed out the railroad and inundated farms. As the floods carried trees and debris downriver, a vast and impenetrable dam of logs formed where the river skirted the Muckleshoot Indian Reservation. The floodwaters were forced back, rising at the confluence of the Stuck River—a small tributary running parallel to the White River for a couple of miles before turning south. In recent years, Pierce and King County farmers had escalated a decades-old practice of forcing flows from the White to the Stuck and back during high water. In flood stage, the White River was a destructive force that nobody wanted.[45]

Several years earlier, a logjam near the confluence had caused water from the White to jump the narrow isthmus between the two rivers and overflow into the Stuck Valley. In an effort to clear the jam, a group of southern farmers detonated dynamite at the base of the bluff. The explosion brought the bluff down on top of the pile instead, further obstructing the river. More of the White River's floodwater was forced down the Stuck channel, carving a path across the narrow barrier between the rivers and splitting the river's flow between the two channels.[46]

In the flood of 1906, this natural barrier fell altogether. The new logjam below the Muckleshoot plateau caused high volumes of water coming downriver to blast a channel through the remaining strip of land separating the White and Stuck Rivers. The floodwater overwhelmed the narrow Stuck channel and scoured the valley as it barreled south toward the Puyallup River and Tacoma. The White River had carved itself a new course, altering the future of King County.[47]

The river that had plagued the farmers was now gone. Along its course were several miles of dry riverbed and a promise of no more floods, ever. The guards deployed by King County were a temporary measure to prevent any attempt to reverse the changes. Pierce County sued. After nearly a decade of legal wrangling, Pierce County gave up the fight in return for its northern neighbor's promise to pay 60 percent of its flood-control costs for ninety-nine years. To the satisfaction of King County farmers, the Duwamish watershed had lost about 480 square miles of its drainage area, taming the violence of its floods. But the reprieve from floods also brought competition for the farmers. The former White River Valley was now eyed by speculators who saw it as ripe for urban and industrial development.[48]

The rerouting of the White River was not, in the end, sufficient to relieve flooding elsewhere in the watershed. The lower Duwamish Valley still received flows from the Green and Black Rivers, where many of the earliest pioneers had settled. The Black, in turn, drained the expansive Lakes Washington and Sammamish. It was also inundated each year with melting snow and spring rains cascading down the Cedar River. By the time King and Pierce Counties had convened a White River Commission and formulated a plan to manage the effects of the river's diversion on Pierce County, King County officials were already designing another diversion to control flooding from the Cedar, directing it away from its confluence with the Black River.[49]

A few years earlier, Green Lake, a small lake in north Seattle with its source in a spring called Liq'tad ("red mud"), had attracted the attention of John Charles Olmsted. A nephew of Frederick Law Olmsted, the designer of New York City's Central Park, Olmsted had been commissioned to create a network of parks for the Seattle area. He proposed a twenty-mile long chain of green spaces and connecting boulevards. Green Lake was considered a jewel in the Seattle parks plan, but residential development had expanded right up to its shoreline.[50]

Olmsted wanted to lower the lake to reclaim land for a perimeter park and trail. Water from the lake was pumped out, dropping the lake level seven feet and reclaiming a hundred acres of previously submerged land. The exposed shallows at the north and south ends of the lake were turned into sports fields. The park was dedicated in 1911. A small creek flowing out of Green Lake through the town of Ravenna to Lake Washington ran dry when the lake dropped and left the lake stagnant. But the practice of lowering local lakes was soon repeated, on a larger scale, to solve flooding problems farther to the south.[51]

The same year that Ravenna Creek was cleaved from Green Lake, the Cedar River was rerouted. A new canal was built to carry the river into Lake Washington, bypassing Renton and alleviating flooding in the town center. The water ultimately drained back out to the Black River, but the remaining link between the Cedar and Black Rivers via Lake Washington would be severed soon enough.[52]

By this time, the construction of the Government Canal championed by Burke was under way. When it opened in 1916, the level of Lake Washington dropped and, as with Green Lake, permanently severed the connection between the lake and its former outlet, the Black River. Lake Washington was not left landlocked as Green Lake had been, however. The rerouted Cedar River and all the waters of Lake Washington and Lake Sammamish now drained out through Lake Union and the Ship Canal into Puget Sound.

Without a water source, the Black River ran dry. The Duwamish leader Joseph Moses described the scene: "That was quite a day for the white people at least. The waters just went down, down, until our landing and canoes stood dry and there was no Black River at all. There were pools, of course, and the struggling fish trapped in them. People came from miles around, laughing and hollering and stuffing fish into gunny sacks." The remaining Native families living near Sbabadid were displaced by the loss of their river and its life-sustaining resources. The Duwamish watershed permanently lost another six hundred-plus square miles, and the Black River ceased to exist.[53]

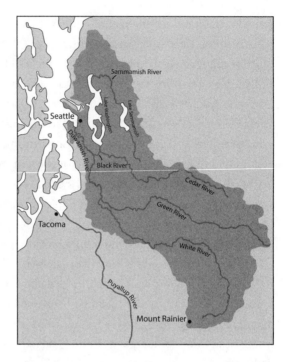

Original watershed of the Duwamish River.

With the White River now flowing south to Tacoma, and the lakes—along with the Cedar River—diverted north through the Ship Canal, only the Green River remained to feed the Duwamish. The Duwamish and the Green Rivers were effectively one. Today the change of name from Green to Duwamish at the former junction of the Black River is the only reminder that several rivers once merged to become the Duwamish. Once extending over 1,600 square miles, today the Duwamish watershed is a mere 480 square miles.[54]

With the opening of the canal, the watershed's extensive network was reduced to a single, tamable river. The Seattle and Lake Washington Waterway Company continued to fill the tide flats at the Duwamish River's mouth. Their contractor, the Puget Sound Bridge and Dredge Company, completed construction of Harbor Island, a three-hundred-acre island in the middle of the tide flats, in 1909. The largest man-made island in the world at that time, it was constructed from fill dug out from the twin canals on its east and west flanks, as well as soil from Beacon Hill that was sluiced

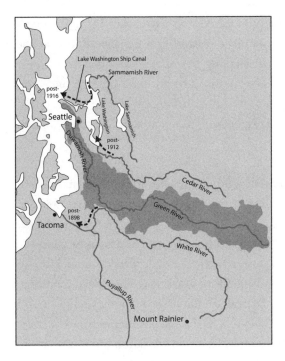

Current watershed. Arrows indicate where the White River, Cedar River, and Lakes Washington and Sammamish were diverted in 1898, 1912, and 1916, respectively. The diversions eliminated the Black River altogether and reduced the size of the Duwamish watershed from 1,600 square miles to 480, leaving only the Green River flowing into the Duwamish today. Courtesy of Tom Sackett, adapted from Puget Sound River History Project.

from the aborted South Canal and the regrading of its northern flank. New industries began to occupy the created lands west of the hill and on Harbor Island, accessed by a bridge over Semple's drainage canals.[55]

The new Duwamish tide-flat industries were seen as a boon to the growing city and elevated Seattle's national stature. One of the first factories on the new island, which opened in 1912, was the Fisher Flouring Mill, one of a handful of wheat mills in Seattle (including the Centennial Mill built on Semple's tidelands). In 1916, the Seattle historian Clarence Bagley boasted, "Today Eastern Washington wheat is manufactured into flour in a Seattle tide water mill before it resumes its journey to the hungry of the world."[56]

A pair of shipbuilding companies that had outgrown their original locations on the downtown waterfront moved to Harbor Island within a few years. Todd Shipyards bought the historic Moran Brothers Company, founded in 1882. They moved the company to Harbor Island in 1918, joining the Puget Sound Bridge and Dredge Company, which later became Lockheed Shipyard.[57]

Todd Shipyard eventually hired Myron Overacker Jr., Quio-litza's grandson, who lived on his family's Beacon Hill homestead. Just as the city's earliest timber and hop farming businesses had hired Duwamish workers, the factories of the early 1900s often hired their second- and third-generation descendants. Along with the city, Native labor shifted from extraction industries to agriculture to manufacturing, continuing to fuel the rise of Seattle.[58]

The industrial boom promoted by filling the Duwamish tide flats encouraged new schemes for the rest of the river's path through the suburban areas south of Seattle. Rail and trolley lines were laid over the former tide flats, connecting the river's early settlements to downtown Seattle. The 1911 Plan of Seattle included a straightened and deepened channel for the Duwamish, an idea that had been proposed in 1892 but judged by county engineers to be "entirely impractical." In the succeeding years, a Duwamish canal was planned with the help of a Scandinavian engineer, and Seattle's chief engineer, Reginald Thomson, traveled to Europe to study canal-building methods. The state legislature authorized the creation of canal districts, and in 1910, south Seattle voters approved a bond measure to establish the Duwamish Waterway Commission.[59]

One member of the newly convened commission was the German immigrant Dietrich Hamm, who had arrived in the United States in 1876 and made his way to Seattle. He opened a coffee house in Pioneer Square, but it burned to the ground with most of the original downtown in the Great Seattle Fire of 1889. After the fire, Hamm and a partner opened a hotel in one of the first brick and stone buildings built in Seattle, a six-story edifice at the corner of James Street and Second Avenue. The establishment thrived during the 1897 gold rush and hosted presidents Grover Cleveland, William McKinley, and later, Theodore Roosevelt. Hamm ran the hotel restaurant; his descendants emphasize his reputation for generosity, which included providing meals to Se'alth's daughter, Angeline, in her later years.[60]

Hamm retained his taste for German-style beer. He did not like the local Rainier beer and resisted the company's efforts to advertise at his hotel. Instead, he shipped Olympia beer from central Washington to serve his guests. According to the Hamm family, Rainier retaliated by jumping Hamm's lease and taking over the hotel. It became a brothel and speakeasy during Prohibition and closed in 1933, the year Prohibition ended.[61]

After losing the hotel, Hamm moved to two hundred acres of farmland in the outpost of South Park. The farm enabled him to support his family and to lease surplus land to newly arrived Japanese and Italian farmers. He got involved in promoting the straightening of the Duwamish River and creating the Duwamish Waterway Commission.[62]

A small stream ran through the farm, draining into the Duwamish River; it would come to be known as Hamm's Creek. Hamm's son described his experience of hiking through the creek's canyon. "It was so beautiful with lots of trilliums and smelling leaves," he wrote. Starting at the spot where a bridge spanned the canyon, he recalled working his way downstream with a fishing pole "until we had 100 fish, all a minimum of 3" and a few larger . . . the fish were good eating."[63]

The Plan of Seattle envisioned all the level ground of the Duwamish Valley from Harbor Island to the Black River (about twenty miles of riverfront) as prime for development. Only the river itself stood in the way of the valley's industrial future: "One of the greatest obstacles in the way of its development is the winding course of the Duwamish River, which swings from side to side." The solution was clear to the engineers of the day: "Further development would properly take the form of simply straightening the river." They saw a better use for the land interrupted by the meanders and its pretty, but pesky, tributary creeks.[64]

Hamm and the waterway commissioners imposed a tax assessment on the river's property owners to pay for the construction of the canal, which terminated at his farm. As one of the valley's largest property owners, Hamm reportedly paid one dollar of every sixteen of the total amount assessed, amassing a significant yearly tax bill. When asked why he took on this financial burden, his granddaughter, Elsa Bowman, said, "I think that he thought it was his obligation, as an immigrant."[65]

Hamm did not live to see the project finished. After his death from whooping cough in 1918, his widow, Aline, had to sell the land to relieve herself of the tax burden. She moved into a modest house on the neighboring property, where Bowman later grew up.[66]

Born in 1933, Bowman remembers her grandmother as a gregarious woman. Aline Hamm was a suffragette, "with her hair cut like Gertrude Stein." She says her grandmother always had a kitchen full of people in the summer, despite having little money. Bowman remembers sitting on the bridge where her grandfather had once fished and watching salmon run up Hamm's Creek. Reflecting on the grandfather she never met, Bowman later wrote, "His leadership of this [canal] project and his service as an elected Duwamish Waterway Commissioner were perhaps his greatest gifts to his adopted city."[67]

Hamm's fellow commissioners shared this sentiment. In a published tribute, they wrote: "Dietrich Hamm during his earthly life erected a monument that is as eternal as the hills. Duwamish is his monument." The sentiment was widely shared as the straightened waterway neared completion. In a letter to Hamm's son, his friend and attorney John Shorett wrote, "When I look down through the great Duwamish Valley everything answers me that he is not dead but still lives, for the work he did will go on and on."[68]

The waterway was designed as a shipping corridor paralleled by two "marginal roads" to the east and west, with railroad lines strategically running between the roads and the canal. The river's former meanders were filled to create new land for industry. It was planned to be three hundred feet wide and to extend past the former Black River, terminating at the historic Green River confluence near Auburn. There was even talk of extending the canal as far south as Tacoma, thus linking Elliott Bay with Tacoma's Commencement Bay via an inland waterway. In 1913, work began on the first phase of the project—a canal reaching from the tip of Harbor Island to a turning basin on a bend five miles upriver. The end of the canal at the edge of Hamm's farm was about halfway to the old Black River junction.[69]

CHALLENGERS AND BOOSTERS

In May 1913, Dietrich Hamm's twelve-year-old daughter christened the newly constructed dredge *Duwamish* by breaking the customary bottle of wine across its bow. Boasting a twenty-four-foot slurry pipe that could pump four hundred cubic meters of mud per hour, it was expected to dredge seven million cubic yards of bottom mud as it sliced a straight channel through the valley. The dredged material would fill the former river meanders and other low-lying areas to create an industrial district for Seattle. On October 13, the *Duwamish* drew its first load of mud from the river bottom.[70]

The machine began digging out one of three planned turning basins, where large ships could maneuver in and out of the waterway. It pumped the slurry onto twenty acres of low-lying land at the King County Poor Farm, where indigent people were housed. Earlier in the year, a federal court had cleared King County's title to the land, which had been seized from the estate of the pioneer settler John Thompson after his death in 1869. The mud generated by the dredging would finance the canal with the sale of fill to the waterfront landowners who sought to develop, lease, or sell their newly created industrial lands. As the dredging and filling began, the value of these future lands was assessed at half a million dollars.[71]

A crowd formed to watch the first mud being removed from the future canal, hear an array of officials applaud its progress, and tour of one of the industries that represented the future of the Duwamish Valley—the new Seattle-Astoria Iron Works plant, which had just moved in near the Poor Farm. A few days later, the *Seattle Daily Times* ran an editorial contrasting the new waterway with Semple's defeated canal. "No sooner is it laid away in the cemetery of the chimerical than the 'Duwamish Project' bursts into full life and being," the paper's editors proclaimed. "It responds to the righteous demand of the South End for a waterway of its own."[72]

Four years later, a four-and-a-half-mile canal had replaced nine miles of winding river bends, but not without immense challenges and a few missteps along the way. Before straightening the Duwamish River, the Waterway Commission had to overcome a barrage of lawsuits challenging the project. Most were claims for damages caused by right-of-way con-demnations and by depriving property owners of their former river front-age. The majority of Duwamish Valley landowners had voted to tax themselves for the construction, but several contested the fees charged for the "privilege" of losing their access to water. The majority of these cases were consolidated into one lawsuit. The aggrieved riverfront landowners lost the case and appealed to the state Supreme Court in 1913.[73]

Samuel Loeb and Benjamin Moyses, owners of the Duwamish Valley's Independent Brewing Company, had property on the river's Georgetown reach—a sharp oxbow that created a bulge of land surrounded by water on three sides. They planned to develop the property as an amusement park. The river was central to the park's design, which was based on Coney Island's Luna Park. But the plan for the waterway would place a barrier upriver of the oxbow, diverting water away from the river bend and leaving the planned park half a mile away from the new waterway.[74]

Loeb and Moyses argued that they had a right to use the navigable river on which they had bought their land. Other plaintiffs included the Seattle Electric Company, which operated a steam plant to power the city's street-cars. The plant was located on a river bend that would also be left high and dry by the diversion. The city argued that the loss of their water intake would require construction of pipelines and a pump station costing $444,000. The Puget Mill Company and other appellants did not claim damages but fought the tax assessment for the work based on the value of the "improvement," rather than on the project's actual cost, which was significantly lower.[75]

At the end of December 1913, the state Supreme Court upheld the decision against all plaintiffs. They ruled that the Waterway Commission was not taking any actual property, since the state was the owner of the riverbeds. Thus "riparian owners are not . . . entitled to damages from the fact that the state will divert the course of the stream and leave their property without access to the water."[76]

After clearing these legal hurdles, the Waterway Commission was free to plow ahead. In early 1915, boosters painted the pending completion of the canal as "the dawn of a wonderful era of industrial development" that would awaken the South End. By March, two-thirds of the canal had been dredged and a section of the waterway opened to shipping. The Seattle Yacht Club hosted a celebratory trip to South Park, about four and a half miles from Elliott Bay. "For eighteen months the big dredge *Duwamish* has been eating out the earth to form the Duwamish Water-way," the *Seattle Daily Times* crowed. "This means that the Duwamish Waterway henceforth is a reality [and] splendid facilities for industrial and commercial expansion are open to immediate development." With

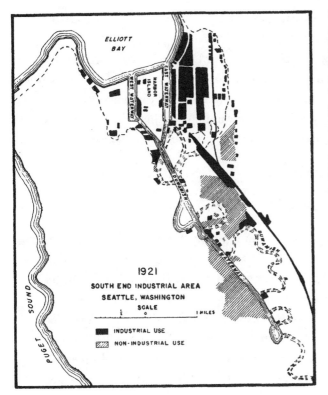

Original (dotted lines) and straightened course of the lower Duwamish River in 1921, with industrial lands marked. From Lucile Carlson, "Duwamish River: Its Place in the Seattle Industrial Plan," *Economic Geography* 26, no. 2 (April 1950). Reprinted by permission of Taylor and Francis Ltd, www.tandf online.com.

only two more miles of the waterway remaining to be constructed beyond South Park, completion was expected within a year.[77]

In January 1916, work had finally begun on the delayed oxbow fill when the *Seattle Daily Times* announced that the dredger was disabled and would be out of commission for at least ten days. A month later, the paper lambasted the dredging company for its slow progress, lamenting that "a great number of the enterprising citizens of the South End who have seen the gigantic task of changing the course of the river begun, will never live to see that feat of engineering perfected if the present 'go-as-you-please' method continues."[78]

It was two more years before most of the promised fill of the river meanders, roadways, and factory sites was finished. As the meanders were being filled, the Waterway Commission also added plans to deepen the waterway to thirty feet to accommodate the newest oceangoing ships. The straightened and deepened canal, extending south from Harbor Island past the city limits, was finally completed in 1920.[79]

As a result of the canal project, the lower Duwamish Valley gained 786 acres of land, 66 of them in the farming community of South Park. Incorporated in 1902, South Park voted for annexation to Seattle five years later. The neighborhood's Italian and Japanese immigrants initially used the new land for farming, but industrialization pressures soon put a squeeze on the valley's agriculture.

INDUSTRY ASCENDING

The Duwamish Valley's oldest and arguably most famous industry is beer, established when the valley was still dominated by agriculture. One of the valley's earliest producers partnered with Eugene Semple in his Beacon Hill canal scheme. Granting Semple access through their swampy waterfront property at the base of Beacon Hill, Bay View Brewing traded its right-of-way for solid fill created by Semple's big dig. Bay View joined the valley's first brewery, Claussen-Sweeney, which opened in Georgetown in 1883 and later became Rainier. Albert Braun Brewing Association joined the field in 1890, near the present-day Boeing Field.[80]

Brewing and hop farming were mutually supportive enterprises in the valley. Farmers grew hops in large commercial plots to supply the growing beer industry, employing first Native labor, then Japanese and Chinese immigrants, to harvest them. As the fields grew in size and expanded upriver, however, aphids began to attack the crops. A devastating infestation, coupled with an economic depression, killed the Duwamish and White River hop industry in 1893. Dairy farms and vegetable truck farms continued to operate in the upper river valleys well into the twentieth century, but along the lower Duwamish River, the former agricultural valley gradually gave way to industry.[81]

On the afternoon of June 6, 1889, John Back was preparing a pot of glue in his boss's woodworking shop at Front Street (now 1st Avenue) and Madison. The pot boiled over, spilling onto the floor and igniting the turpentine and wood shavings covering the shop floor. Back tried to douse the flames with water, but the liquid caused the turpentine to spread. The flames soon engulfed a neighboring liquor store and two saloons, adding alcoholic fuel to the fire. Fire brigades responded, but the city's private water supply could not provide adequate water pressure to fight the flames, and several of the wooden water mains were themselves consumed by fire. By 3 a.m., flames had consumed all of downtown Seattle and the waterfront wharves. The skeletal remains of some brick frames and chimneys were all that still stood.[82]

The following day, Mayor Robert Moran—who had lost his own shipyard in the blaze—met with six hundred business owners to revamp the city's fire code. Before 1890, Seattle had been built out of the timber surrounding the city. The new rules required that all buildings be built of brick. The pace of rebuilding was frenetic: by some accounts, 465 brick buildings went up in a single year following the fire.[83]

John McAllister was an inadvertent beneficiary of the fire's devastation. As one of Seattle's few brick makers, McAllister had recently moved and expanded his Lake Union company onto the Duwamish tide flats. He soon needed to double production to meet the new demand caused by the fire. Within a few years he moved again, opening McAllister's Brickworks in South Park. At the time of the move, another brick factory was already operating across the river in Georgetown.[84]

Within months of the blaze, the Puget Sound Fire Clay Company had opened their brick factory a few miles upriver of Seattle. The factory soon dominated a portion of the old Van Asselt land claim on the Duwamish riverbank. The pioneer settler Arthur Denny bought the plant and its associated clay mine in Renton. His renamed Denny Clay Company specialized in making clay pipes to replace the city's old wooden water and sewer lines. He opened a new plant and mine in Taylor, Washington, and partnered with the Renton Clay Company to expand their mines, rebranding as the Denny-Renton Clay and Coal Company. Along with coal mining, clay became big business in the upper valleys of the Duwamish watershed, and the new brick factories joined the breweries in the Duwamish Valley.[85]

The influx of industry attracted new immigrants to the southern riverfront suburbs. The breweries built company housing, and new

Turn-of-the-century advertisement for the Duwamish Valley's Denny Clay Company. Printed in Seattle Fire Department Relief Association, *Twentieth Century Souvenir of the Fire Department of Seattle* (1901).

working-class neighborhoods grew up for the brickworkers. An influx of German and Eastern European workers joined the valley's Italian and Asian farming families. By 1910, South Park and Georgetown had voted for annexation to Seattle, further integrating their future with that of the urban and industrial financiers who were buying vast swaths of farmland as the city rose from the ashes of the fire.[86]

Two more industries were also emerging in early-twentieth-century Seattle. Forges and blacksmiths' shops employed growing numbers of German immigrants. The fourth-generation resident Marianne Clark recounts the arrival of her great-grandfather, Albert Premel, in the early 1900s. Born in 1853, Premel defected from a German navy ship in New York in 1872 and eventually made his way to Seattle, settling on the oxbow fronting Georgetown's Charleston Street (now Corson). He worked as a blacksmith, possibly at the stables for the nearby racetrack. He later moved to

William Boeing's yacht, moored at the Heath Shipyard on the Duwamish River, near the current location of the First Avenue Bridge. Boeing commissioned Heath to build the vessel, then purchased the shipyard in 1910 and converted it to the Boeing Company to build airplanes in 1917. Courtesy of the Boeing Company.

neighboring Carleton Street, where Marianne and her husband raised their own daughter and still live today.[87]

Each of these early industries left their mark on the landscape and character of the Duwamish Valley. The next business to rise from the tide flats, and the first to be built on the river itself, proved to be transformative—not just for the Duwamish River, but for the city as a whole. Edward Heath built the river's first shipyard, relocating from Tacoma to a swampy meander a couple miles upriver of Harbor Island. In 1909, he began building wooden ships in a large barn built on top of two hundred pilings sunk into the marshy ground.

One of Heath's first commissions came from a young timber industry magnate. The well-heeled and well-connected William Boeing had arrived in Seattle just the year before. When the shipyard that was building his yacht found itself struggling in 1910, the young Boeing bought the company for ten dollars, rescuing it from bankruptcy and relieving Heath of his debts.[88]

On the opposite bank of the marshy river, just south of Denny's brick-works, Seattle was treated to its first flying show above the Meadows Race Track in March 1910. The Meadows consisted of a whitewashed hotel, a boardwalk, and a mile-long circular track, but horse racing had been shut down by changes in the law several years earlier. Twenty thousand people, many riding open railcars traveling from Seattle, arrived at the Meadows to watch the aerial acrobatics of the "Crazy Man of the Air," Charles Hamilton, flying a Curtis biplane. Hamilton briefly wowed the crowd before his plane crashed into standing water in the center of the track, which frequently flooded during high tides.[89]

After the exhibition, William Boeing traveled to Los Angeles to watch the nation's first airplane races. He was smitten with the dream of flying but had to wait another five years before he had the chance to experience it himself. In 1915, the aerial performer Terah Maroney visited Seattle. Boeing and a friend lined up along with other enthusiasts and joined Maroney on several of his flights. Captivated by the experience, they soon set to work building their own seaplane, convinced that they could improve on the design of Maroney's plane. In 1916, they founded the Pacific Aero Products Company, and in 1917 Boeing converted the Heath shipyard into the Boeing Airplane Company. Under the guidance of his company's first engineer, Chinese-born Wong Tsu, Boeing designed and sold fifty C-2 float planes to the US military, just in time for America's entry into World War I.[90]

Around the Boeing factory, the river was changing. Between 1915 and 1917, dredgers straightened that section of the Duwamish. The cut imposed a deep, straight canal through the river's bends and twists. Boeing's expand-ing plant sat above the marshy tide flats between the shipping canal and the river's old meander until infilling with dredge spoils and waste mate-rials created a mound of emergent land. Dubbed Foss Island, the offshore mound grew until the channel separating it from the mainland was com-pletely filled in the late 1960s.[91]

Boeing's work building military aircraft expanded into the production of mail planes for the US Postal Service in the 1920s. A small airfield was cleared about a mile north of his plant, on a level field next to a remnant river bend. The company flew its mail carriers from the new field and soon built its first passenger airplane, which took its maiden flight on February 8, 1933. They showcased the marvel at the Chicago World's Fair later that year. Business boomed, and the company's need for both manufacturing and airfield space grew.[92]

The Meadows racetrack across the river, where Boeing saw his first flight, was soon transformed into the King County International Airport to support the growth of the aviation industry. County residents taxed themselves $950,000 for its construction, and fifty thousand people attended the 1928 inauguration of the valley's new airport, nicknamed Boeing Field. Boeing sent a letter thanking the King County Commissioners for their initiative in building the airfield. "I am deeply sensitive of the great honor," he wrote. "I believe that flight and air transportation are going to take a more important part in our civic and national progress than we are able to foresee today." With a state-of-the-art airport at his disposal, Boeing just needed a bigger factory to keep expanding his company.[93]

One of the most prominent of South Park's farmers was Giuseppe Desimone, who championed the creation of the Pike Place Market in 1907. Desimone and his neighbors navigated the Duwamish River's twists and turns to float their produce to the market or transported it on truck wagons over a tide-flat trestle road. Seventy-five percent of the market's stalls in the early years were occupied by Japanese fruit growers from the White River; most of the rest were Italian vegetable farmers located along the lower Duwamish River. Desimone bought and drained swamplands in the former river bends to create more farmland. In time, he was able to buy shares in the Pike Place Market and eventually became the market's majority stockholder.[94]

Desimone was committed to making the market produce affordable and accessible, keeping costs low and ensuring the survival of the valley's working farms. However, his largest contribution to his adopted city may inadvertently have helped to push agriculture out of the lower Duwamish Valley rather than preserve it.

Desimone went from renting farmland in Georgetown to owning three hundred acres, much of it west of the river in South Park. After the river was straightened, some of Desimone's lands were left on the east side of the canal, next to Boeing Field. His granddaughter Suzanne

Hittman recalls growing up on the South Park farm. "We lived right on the city line," she remembers. "It was very pastoral." Their neighbors were other Italian farmers and a German family who raised chickens. Unlike most of the other local families, her grandfather owned additional farm-land, growing beans along the river in the rich soil of the floodplain, which was so prolific that he sold most of the produce wholesale to canneries in Kent.[95]

In the 1930s, Boeing was looking for a new place to build a bigger factory and started talking about moving to California. Like the rest of the nation, Seattle was suffering from the effects of the Depression, and it had seen its unemployment rate grow from 11 percent to 25 percent in the early 1930s. The city was desperate for jobs. The former tide flats at the mouth of the river had been turned into a makeshift encampment of mostly homeless men, who named it Hooverville after President Herbert Hoover. This was land once occupied by the Centennial Flour Mill—the land that Eugene Semple had bragged about creating in 1897. Hittman credits her grand-father's attorney, Harold Scheffelman, with convincing Desimone to give Boeing a parcel of property on the river, with 250 feet of waterfront and enough space for a large plant, to keep the prosperous company in town.[96]

The now-famous story is that Desimone sold the Boeing Company the land for their new factory for one dollar. But, insists Hittman, "My grand-father wouldn't have gifted it without [Scheffelman's] influence," noting that Desimone was a real-estate investor in the business of making money. A young attorney who later worked with Scheffelman agrees. "I don't think it was Giuseppe Desimone saying, 'Oh, I love Boeing,'" said David Sweeney in 2019. "I suspect that he was negotiating to sell Boeing some land, and thought that if Desimone gifted the property, Boeing would be likely to buy more land from him elsewhere in the area."[97]

In 1936, Desimone sold—or gave—Boeing the land the company needed to expand its airplane factory. Boeing initially built the expansion, known as Plant 2, on these twenty-eight acres of former farmland. The facility later expanded to include more than thirty industrial and office buildings on forty acres of land south of the King County Airport. During World War II, the new factory became the nation's primary warplane production factory, churning out up to seventeen B-17 bombers every day.

Employment at the plant increased from 7,500 in 1940 to 46,000 by the end of the war.[98]

By 1945, the lower Duwamish Valley was filled with industrial development. Many of these early industries would later become the focus of environmental cleanup efforts. The canal provided not only access to water but also cheap and easy disposal of waste. The industrial boom introduced new kinds of chemical by-products that had to be disposed of—legally or otherwise.

OF FISH AND DAMS

From 1920 to 1945, industrial businesses claimed 1,270 acres of land in the Duwamish Valley, and city planners were becoming concerned about the shortage of additional industrial land. Seattle's City Planning Commission called for a study of potential new industrial lands in the upper Duwamish and Green River Valleys. The study's engineer, Lars Langloe, observed: "The topography and transportation systems of the city and its environs do not invite such expansion in any other direction." To facilitate industrial growth upriver, Langloe recommended that the constructed Duwamish Waterway be extended southward.[99]

While land was still available upriver, it was too swampy and flood-prone to be of use to industry in its natural state. Moreover, in Langloe's view, the river's curves presented "a winding obstruction to orderly progress." He warned that industry would leave Seattle if industrial land were not made available quickly. "The river must be regulated now," he concluded. He proposed a five-and-a-half-mile extension of the Duwamish canal from Tukwila to Auburn, straightening and shortening the course of the upper river by nearly three miles. He estimated that the project would create 920 acres of new industrial land. He also encouraged the construction of a storage dam to control floods in the Green River Valley–an idea that was already under review by the Army Corps of Engineers but one that, like the Ship Canal in earlier years, they did not yet consider financially justified.[100]

The economic geographer Lucille Carlson seconded Langloe's call for a Green River dam in 1950. She estimated that "if made flood free, approximately 20,000 acres of land will be safe and available for industrial and agricultural use." She also supported Langloe's recommendation for extending the canal, warning, "Industry must either crowd out non-industrial users or move still farther south along the Duwamish."[101]

The canal extension proposal failed to secure the required regional consensus. Carlson and Langloe's endorsement of a Green River flood-control dam, however, added momentum to that proposal. A decade before Langloe's report, the Army Corps had recommended a dam as the best way to alleviate the flooding that continued to plague local farms, but the proposal was opposed by fishing interests. The Cedar River had already been dammed to form a drinking-water reservoir. The barrier cut off salmon spawning grounds and was believed to be partly responsible for the local fishery's decline. The objections of fishermen to another dam in the watershed caused the idea of a Green River dam to be shelved with the approach of World War II.[102]

Pressure on Puget Sound's fish stocks, including the salmon runs of the Duwamish River, had been increasing in the decades leading up to the war. Native people had always harvested salmon for trade, as well as to preserve and eat through the winter months, but the market for Puget Sound's fish was initially slow to grow. Salmon fishing and canning operations were centered on the Columbia River. But as the Columbia River fishery started to show signs of strain, the fishing industry began to look to Puget Sound. The inland sea's first cannery was built in 1877 at Mukilteo.[103]

Puget Sound salmon exports grew from 5,000 cases in the first year to 90,000 cases in 1893 and 2.5 million in 1913. The value of salmon packed by Puget Sound's forty-five canneries was estimated at $16 million, with 90 percent of Washington State's salmon exports coming from the Nisqually, Puyallup, Duwamish, Skagit and other rivers flowing into the sound.[104]

Seventy years after white settlement, the fishing rights explicitly granted to local tribes in the treaties had been usurped by commercial fishing interests, and fish stocks were in steep decline. In 1900, the year the upper Cedar River was dammed, concern about declining fish runs spurred the State Fish Commission to construct a salmon hatchery on Soos Creek, off the Green River, on the site of a former Native fishing village. In the 1920s, the State Fish Commission declared a "fish crisis" in Washington State. Because the Soos Creek Hatchery was not succeeding in restocking the Duwamish, the state closed the river and bay to commercial fishing altogether.[105]

As governor, Eugene Semple had tried to pass fishing regulations in Puget Sound when the Columbia River fishery began to show signs of decline in the 1890s, but he was rebuffed by the state legislature. By the 1920s, state lawmakers were more amenable to imposing restrictions—especially on Native people. Tribal fishermen had long been easy targets for a colonial government that wanted Native people, and their resource use, confined to the reservations. Fishing spears and snares used mostly by the tribes were prohibited in the 1910s, and officials began to arrest Native people found fishing outside their reservations. One Muckleshoot resident complained of being taken prisoner by a game warden while fishing on the Green River and jailed for two days.[106]

The treaty guarantees that the tribes could continue to hunt and fish throughout their "usual and accustomed areas" were considered irrelevant. Commercial fishing took the lion's share of salmon caught in Puget Sound. Much of it was done with traps made of wood and nets—a technology originating with the Puget Sound tribes that had been adapted, with great success, by the region's new commercial fishermen. A single trap could yield a harvest of up to sixty thousand salmon in a single day.[107]

Fishermen who ran gill nets and purse seines from boats competed with the fish trappers. As competition intensified, each interest group tried to pass regulations that would restrict their competitors' share of fish, but with little success. The game changed, however, with the rise of sport fishing in the 1920s. Sports fishermen were largely white and middle-class, with more wealth and political power than the newer immigrant fishermen who dominated the commercial salmon industry.[108]

In 1934, the Washington State Sportsmen's Council, representing forty thousand recreational fishers, allied themselves with gill netters, purse seiners, and trollers—who did much of their fishing in the far-off Strait of Juan de Fuca and northern islands—to lobby for a ban on fixed-gear fish traps. They passed an initiative prohibiting "any pound net, fish trap, fish wheel, scow fish wheel, set net, weir, or any fixed appliance for the purpose of catching salmon, salmon trout, or steel head." As a result, in most of Puget Sound, sport anglers had the region's remaining salmon runs to themselves.[109]

With proposals for a Green River dam to mitigate flooding in the Green-Duwamish Valley, fishermen worried that it would spell the end of their struggling salmon runs. With the fishery in decline, large numbers of unemployed immigrant and Native fishermen went to work in the valley's factories, which were booming with the war effort. After the war, when they tried to return to fishing, they found the regulatory and political landscape changed, much to their disadvantage.

In postwar Puget Sound, farmers who had remained in the Duwamish Valley resumed their lobbying for flood control. Supported by Langloe's report, Congress finally approved the Green River dam in 1950. The dam was dedicated on May 12, 1962, in the presence of thousands of valley residents who traveled to the mountainous upriver site "to witness the beginning of their flood free days." The remaining "wild" reach of the Duwamish watershed was tamed by the dam, reducing peak flows in the river by nearly 75 percent.[110]

Ironically, the building of the dam ultimately led to the death of the vast majority of farms in the Green River Valley. As Langloe had predicted, controlling floods on the Green River opened the valley to industrial development, crowding out farmers. By 1979, the Green River Valley had lost twenty thousand acres of farmland to industrial and urban expansion. The expansion of industry led to the creation of dense residential and commercial centers for workers and their families. Fifty years after the dam prevented its first flood, the local historian Alan Stein commented on the changes. "Now instead of engorged waterways, the valley is filled with glutted highways and streets," he noted. "Traffic jams have taken the place of logjams."[111]

ETHNIC CLEANSING—AGAIN

Fishermen and farmers were not the only Duwamish Valley residents who found their lives upended by the war. On December 7, 1941, Japan bombed the US military base in Hawaii's Pearl Harbor, leading the United States to enter the war. Within hours of the attack, the US government rounded up more than five thousand Japanese community leaders living in the United States. They were already on an FBI watch list of potential spies because of their status as businessmen, teachers, or priests in the Issei (first-generation) Japanese community in America. Many were torn away from their families, which included children who were American citizens. On February 19, 1942, President Franklin D. Roosevelt issued an executive order authorizing the internment of all individuals of Japanese descent living along the US West Coast. Eventually more than seven thousand Japanese Americans from the immediate Seattle area were taken.[112]

After the forced displacement of the region's Native people, the internment was the largest demographic shift ever seen in the Duwamish Valley. The effect was devastating. South Park and the Duwamish-Green River Valley from Seattle through Tukwila and into Auburn were farm country. In the years before the war, the majority of the region's produce was grown on Japanese-owned farms, and Japanese were employed as farmhands on most other farms. Giuseppe Desimone employed Japanese workers on his extensive acreage along the Duwamish River. Suzanne Hittman recalls the sudden disappearance of roughly a quarter of her classmates, and fields empty of farmhands.[113]

"Many of those who worked on the farm never returned," recalled Hittman in 2018. She remembers that Filipino workers took the place of interned Japanese farmhands, and the African American and Latino children of new workers in the Boeing factory replaced her missing friends and classmates.[114]

Census records reflect those changes. Before 1940, South Park's population was 12 percent Japanese. After the war, Japanese residents constituted only 1 percent of the population. In their place, African American and Latino families made up 11 percent and 4 percent of the neighborhood's population, respectively, in the postwar years.[115]

Across the river, on the ridge at the top of Beacon Hill, Quio-litza's grandson, Myron Overacker Jr., was still living on the land homesteaded by his parents, when his neighbor, Tadashi Yamaguchi, was taken. Because Yamaguchi ran a store out of the first level of his house, he was targeted by the FBI as a successful businessman. He was arrested and taken to an army camp in Montana within hours of the Pearl Harbor attack.[116]

Yamaguchi frequently traded favors with Overacker, who was a skilled handyman. Yamaguchi's son, Kay, an American citizen, took over managing the family store after his father was taken, but when the rest of the Yamaguchi family were interned the following May, they asked Overacker to watch over their house.[117]

Seventy-six years later, Tadashi Yamaguchi's grandson, David Yamaguchi, recounted the family's stories of trying to keep their family together during the years of internment. His great-aunt, Catherine Yamaguchi, also known by her Japanese name, Natsuko, corresponded with the Overackers regularly. Writing from Camp Minidoka, Idaho, in May 1943, she said, "We received a letter from the U.S. Department of Justice stating that dad's case had been reconsidered. [H]e can now join us here as soon as a transfer can be accomplished. . . . [A]ll of us are hoping it will be soon." When the family finally returned to Seattle and found their house and shop intact and well cared for, they moved back in and picked up the threads of their life in Seattle.[118]

In an unexpected twist, David Yamaguchi was able to return the kindness of Myron Overacker's Native family two generations later by shedding new light on the local tribes' founding stories. Yamaguchi, a forest scientist in the Pacific Northwest, was traveling in Japan when he heard stories of an "orphan" tsunami—one whose cause was unknown. Research into this event led to the dating of a large earthquake off the coast of Washington in the year 1700. Yamaguchi had previously visited a "ghost" forest of dead trees on the Washington coast. One theory about the cause of its destruction was a massive inundation of seawater, such as a tsunami causes.

Yamaguchi used tree-ring dating to determine the age of the trees, which corresponded perfectly with historical accounts of the Japanese tsunami. As a result, the phenomenon could be dated precisely to January 26, 1700. An earthquake of the magnitude necessary to send a massive tsunami to Japan would have shaken Native villages throughout Puget Sound. Although there are no written records of this period in the Northwest, evidence did survive in the form of local oral histories. The 1700 earthquake is believed to be the basis for the Native legend of the battle between the Thunder God and his supernatural rival.[119]

Most of the interned Japanese families living in the Duwamish Valley below Beacon Hill were less fortunate than the Yamaguchis. When they returned from internment, their farms and homes had been lost to new occupants or were gone altogether. In time, Southeast Asian immigrants took their place. By the end of the twentieth century, South Park had become home to growing numbers of Vietnamese and Cambodian residents, many of whom were attracted to the riverside neighborhood where they could carry on their fishing traditions. Immigrants from other places moved into South Park and Georgetown as well. Latino families arriving in the years after the war also recall fishing in the river, as well as working at Boeing, sweeping refuse and manufacturing by-products into the river below Plant 2.[120]

4

Tears on the Fenceline

The High Cost of Pollution

1950–2000

"EMBATTLED HOUSEWIVES PICKET DUMP," PROCLAIMED A HEAD-line in the *Seattle Daily Times* on November 1, 1961. The reporter, Byron Johnsrud, explained: "The hornet-angry victims of 'garbage-dump fallout' are prepared to string up all city officialdom from the highest clothesline if garbage-burning at the South Park dump is not halted forthwith."[1]

The story followed months of tussling between the residents of South Park and officials at City Hall over noxious smoke and smells from open burning at the south-end city dump. A week after the news story, Reverend Robert Morton of the South Park Methodist Church followed up with his own impassioned letter to the editor. "This flaming dump is a con-tinuous inferno, belching debris upon every structure within a mile area," he complained. "Schools, churches, mercantile establishments and homes are the recipients of smoke, rats, smog and fallout from this man-made hell."[2]

In 1959, the city had rejected calls for an end to the burning, despite receiving a petition from neighbors complaining that the stench was ham-pering development of the surrounding area due to "smoke and attendant conditions." The city's engineering department determined that to stop the

View from the First Avenue Bridge in 2007, showing air pollution from factory smokestacks along the Duwamish River. Courtesy of Paul Joseph Brown.

burning would cost too much, requiring the purchase of a $45,000 bull-dozer to spread out the trash instead. More critically, the increased volume of unburned trash would shorten the life of the landfill from fifteen more years to only three. City departments discussed the possibility of building a modern incinerator, but with the cost estimated at $7.5 million, those discussions petered out. In the meantime, soot continued to rain down on South Park, and neighbors suffered from the smoke and fumes. In 2006, Jackie Jacquemart, a local resident, reflected on her memories of trying to dry her family's clothes at that time: "If you didn't get it off the line fast enough, it was covered in soot and you'd have to wash it again!"[3]

Wearing gas masks and carrying signs reading "Let us breathe" and "Save our laundry!," South Park's housewives picketed at the dump, strung up their soot-blackened laundry for all to see, and staged a sit-in at City Hall, demanding to speak with their elected city council members. After several weeks, the council relented. At the end of November 1961, its Public Safety Committee announced that burning would stop immediately, and that the fire smoldering at the dump would run out of fuel and die a natural death by the end of the year. "This is the Council's Christmas present to South Park," announced Councilman Clarence Massart after the committee vote. The *Seattle Daily Times* published a glowing editorial under the banner headline "When Women Speak Out in Protest." "To the ladies of South Park," the editors wrote, "our congratulations on your achievement."[4]

The victory was a turning point for the embattled neighborhood, but pollution from the dump was not the only threat facing South Park residents. In 1956, Seattle had approved a new comprehensive plan that designated South Park and its surroundings—including the smaller enclave of Georgetown across the river—as an area in "transition to industrial." City planners were preparing for the next stage of industrial expansion in the Duwamish Valley. Rezoning was expected to replace the historic Duwamish Valley neighborhoods with a new industrial landscape, as envisioned by the builders of the Duwamish Waterway decades earlier. The designation put local residents on notice that their days on the river were numbered. In the mid-1960s, under increasing pressure from business interests, the city's planners were ready to rezone the entire Duwamish

Valley as industrial. But South Park's emboldened residents wouldn't be moved so easily; the neighborhood was ready to fight.[5]

For several years, South Park residents had been pressuring the city council to undo the rezoning plan. In 1967, they secured a recommendation from the Seattle Planning Commission to preserve residential zoning for the neighborhood. At a planning committee hearing on the question, a neighborhood resident, Tony Ferrucci, argued that Seattle had treated South Park as "a stepchild" in city development and that the historic neighborhood deserved to be protected. But disruptive opposition from industrial representatives caused the committee to call a recess before the hearing was over. It was more than a year before the issue finally made it out of committee and to the full city council for a vote. "My folks grew the first crops that went into the Pike Place Market," said Ferrucci after the aborted hearing. "Ours is typical of the old South Park families who want to keep on living where [we grew up]. What am I supposed to do, throw my parents in the garbage can?[6]

Jackie Jacquemart, a veteran of the garbage-dump protests, later recalled how anxious she and her neighbors were about the possibility of losing their homes. "We fought like heck," she recalled in 2018. "They wanted all the houses gone." Years later, the *Seattle Post-Intelligencer* claimed that a contingent of South Park's 4,200 residents staged a "march on City Hall" to protest a future industrial zoning for the neighborhood. There are no contemporaneous news reports of a march (though one reporter did ambiguously refer to an "incident" at City Hall), but South Park residents were on hand on February 26, 1968, when the Seattle City Council voted unanimously to amend the city's comprehensive plan to remove the "transition to industrial" designation.

There was one catch: the previous year, local developer John McFarland had requested that six acres of land that he leased be immediately rezoned to industrial. Attempting to appease competing interests, the council agreed to preserve most of South Park as residential but also directed the Seattle Planning Commission to study the possibility of rezoning a corner of the neighborhood, including McFarland's six acres, for industrial use.[7]

Speaking for South Park's residents about the area still in dispute, Ferrucci told Walt Woodward of the *Seattle Times*, "It is not a run-down area.

It has paved streets and sidewalks. Homes are kept up by families in the lower- to middle-income range." But, said Ferrucci, the threat of rezoning was putting financial pressure on homeowners. "We can't get mortgage money with that cloud of uncertainty over us." While waiting for the city council's final decision, Seattle's planning director, John Spaeth, noted, "This is no longer a scientific planning matter, but a political question. Has the time come for industry in South Park?" Spaeth speculated that if industry were allowed a foothold, the rest of South Park might well begin to lose its residential identity as well. In the end, all of South Park, including the disputed six acres, remained residential.[8]

Though South Park residents retained their historic Duwamish Valley neighborhood, and the dump fires of the previous decade had been squelched, the conflict between residents and industry, and the problems of disposing of the wastes they all generated, would continue to smolder.

GARBAGE IN, GARBAGE OUT

As a boy in the 1960s, Greg Wingard often rode his bike down the old logging roads in the woods around his Maple Valley home. One of his favorite places to explore was a hilltop above Rock Creek, near the old mining town of Landsburg. The hill lay on the divide between the Cedar and Green Rivers. A deep, jagged trench ran along the ridgeline, exposing the innards of the mountain and providing hours of entertainment for the young rock hound. A set of plaques at the top of the hill commemorated several miners who were killed in 1955 when the coal tunnels where they were working—an area known as the Rogers seam—caved in, creating the open gorge. As Wingard grew older, he and his friends used the debris they found in the trench for target practice—construction materials, mattresses, and even a few metal drums, which rewarded them with a satisfying ping when hit with their 22-caliber bullets.[9]

In the early 1980s, Wingard returned to the trench as an adult to see if the drums he remembered shooting at were still there. They were. Wingard, then working for Greenpeace, was joined by a friend, Mike Bray, who was a stream walker with the New Jersey Public Interest Research Group.

Bray's job involved investigating and sampling potential sources of toxic pollution. He and Wingard had teamed up to investigate a number of suspect sites near rivers and streams around Seattle.[10]

When Wingard and Bray explored the Rogers seam, they realized that the gorge contained dozens of metal drums. Some were riddled with bullet holes and mostly empty, but others, judging by the range of tones obtained by rapping on the drums, still held contents. "It was really common practice to use southeast King County as a dumping area for a lot of stuff that's pretty nasty," Wingard recalled. He realized he could do something about it. In a 2018 interview, he remembered thinking, "I can raise hell, and hold somebody accountable."[11]

Wingard took photographic evidence of the trench to the regional office of the US Environmental Protection Agency in Seattle and the state Department of Ecology in the early 1980s and hounded the staff until he got a response. He staked out the site and observed ongoing dumping, recording four truckloads of oil-tar waste being dumped into the seam during the summer of 1983. He found allies at the county health department who investigated and levied fines on the property owner, the Palmer Coking Coal Company, which steadfastly denied that any of the material dumped in the trench was toxic waste. "Hell no, we're not fools," insisted a company co-owner, Evan Morris. "We haven't dumped any hazardous materials there." The company acknowledged using the trench to dispose of woody debris and construction materials but denied the existence of the drums that were clearly visible.[12]

The health department eventually succeeded in curbing the use of the mine site for hazardous waste disposal, but once the dumping stopped, the agency moved on. The drums remained in the pit, despite concerns about the effects of chemical waste leaking into local water sources. Wingard, worried about nearby groundwater and drinking wells, continued to push for cleanup of the site. Eventually, he got the ear of Lee Dorrigan, the Washington State Department of Ecology's toxic waste division manager, who assigned a state investigator to check out the hilltop.[13]

Wingard escorted the inspector, Norm Peck, to the site in 1989. Together they scrambled around the edge of the trench as Wingard described his past visits and observations. After a while, he noticed that Peck was no longer listening. "Suddenly, he was like a pointer—his hunting dog mentality kicked in," said Wingard. Peck uncovered at least two hundred drums of waste during that visit, and he suspected that was just the tip of the iceberg. He reported back to Dorrigan, and an official investigation was launched.[14]

By the mid-twentieth century, in addition to running out of room for their factories, the Duwamish Valley's industries were running out of places to dispose of the chemical by-products they generated. The city's engineers had neglected to prepare for this necessary consequence of industrial expansion. "Everyone was at fault," said Dorrigan. "At the time, there was simply not enough hazardous-waste landfill space."[15]

In the late nineteenth century, coal was big business in the Green and Cedar River Valleys. Mining activity peaked between the 1880s and the 1920s. By the 1950s, coal was in decline, and several former company towns had shrunk to a handful of weekend cabins or ceased to exist altogether. The sparsely populated area gradually became the main destination for Duwamish Valley industry wastes, in both permitted and unpermitted landfills.[16]

Near the Cedar River, the Queen City Landfill was a permitted disposal facility used by many factories. One of its biggest clients was the Boeing Company, which dumped tankers of liquid waste from its Duwamish River plant into deep pits at the landfill beginning in 1957. As demand grew and space became tight, problems with waste containment at the landfill began to emerge, and evidence of contamination of several ponds and groundwater at the site came to light. In the mid-1960s, the owners were forced to close the dump and stop accepting new material.[17]

By then, Boeing had shifted its waste disposal to another facility, called Western Processing, but this site, too, was soon facing shutdown over pollution concerns. By that time, Boeing alone had already shipped twenty-four million gallons of industrial waste to the two disposal sites, and its annual volume of waste was on the rise. Waste transport and disposal contractors began to look for new places to take the growing volumes of waste from local businesses. The Rogers seam of the Landsburg mine was pressed into service in a desperate—and illegal—bid to find a place for the Duwamish industrial area's excess hazardous waste.[18]

The mine was owned by the Morris and Kombol families. George Morris had immigrated to the United States from Wales in 1880 and slowly made his way west, working in coal mines along the way. In 1921, his sons purchased the old Durham mine, founded in 1886, one of the first coal mining operations in the Green River Valley. Incorporating as Morris Brothers Coal Mining Company, the family changed the company's name to Palmer Coking Coal in 1933 and later purchased the Landsburg mine property. One of George Morris's granddaughters, Pauline Morris, married Jack Kombol, whose father arrived in Washington State in 1902 to work as a miner, like many other Croatian immigrants. The marriage between the two mining families resulted in a business partnership that continues today. Now in its third generation, William Kombol is one of about forty Morris family co-owners and serves as the current manager of the Palmer Coking Coal Company.[19]

In the 1970s, George Morris Jr., a second-generation co-owner of the company, also owned a roadside cafe near the Landsburg mine. He began charging disposal companies a dollar per barrel to use the defunct trench as a dump. Waste haulers regularly pulled into the parking area at Morris's E-Z Eatin' Cafe, refilled their coffee, deposited cash in a box by the door, and plucked out the keys to the locked gate on the road leading to the Landsburg mine. Undetected, they dumped their loads into the trench, locked the gate behind them, and returned the keys to the box on their way out.[20]

One of the hauling companies that routinely dumped solvents and other wastes from Boeing and other companies into the Landsburg trench was ChemPro. It was based in Seattle's Georgetown neighborhood, just a stone's throw from the Duwamish River. Its owner, Ron West, personally handled much of the waste material at the company's Georgetown transfer and storage yard. The former headquarters of his company is now itself a hazardous waste cleanup site. West died of brain cancer at the age of thirty-nine.[21]

Lloyd Noll, a company driver, recalled the key drop. "Often I had to wait and have another cup of coffee because some other dumper was using the key," he told Danny Westneat, a *Valley Daily News* reporter, in the early

1990s. Noll also told Westneat about the many times he had spilled or been splashed with the chemical waste he was hauling. "I felt unsettled all the time, but it was a job," he said. "Everyone at ChemPro knew we were doing illegal things."[22]

In time, Noll's hands and arms became pocked with warts. Noll himself had called the Environmental Protection Agency to blow the whistle on his employer in 1979, but, he said, nothing was ever done. He continued to pick up wastes at industrial sites ranging from Boeing to the federally operated Hanford Nuclear Reservation, 150 miles away, and carted them into rural King County to dump them into the Landsburg trench.[23]

After reading her investigator's initial report on the Landsburg mine, Lee Dorrigan ordered a helicopter flyover of the site in 1989. From the air, it was easier to observe the size of the exposed trench and estimate the number and distribution of the barrels it contained. The Department of Ecology issued a notice of violation and began a formal site assessment. Eventually, more than 4,500 drums were found buried in the old mining trench, along with roughly two hundred thousand gallons of liquid chemical waste. The trench was nearly vertical, 60 to 100 feet wide, three-quarters of a mile long, and up to 750 feet deep. Where the contaminants dumped into it might end up was unknown.[24]

A few hundred drums were eventually removed from the Landsburg mine in the 1990s. The state's cleanup order included long-term testing of surrounding soil and water to monitor for the chemicals that were found in the drums, which included heavy metals, cyanide, and cancer-causing hydrocarbons. In 2017, chemicals dumped into the trench were finally detected in groundwater nearby. Dioxane, a solvent and suspected carcinogen, was found in several monitoring wells that had been installed in the hillside surrounding the trench.[25]

After hearing about the discovery, Wingard wrote to Lee Dorrigan, who was then retired, about the new findings. "The thing I find disturbing is that 1,4-dioxane is, chemically speaking, the canary in the coal mine." As a water-soluble chemical, he said, dioxane can travel quickly and easily. "Given the amount of time this took to show up, I am concerned that some of the other highly toxic wastes that were dumped may take years more before we notice them migrating."[26]

The hillside containing the old Landsburg mine is currently slated for residential development. A school has been built across the road, and permits have been submitted to build a mixed-use residential community within the gated area once used by miners and waste haulers. Whether the recent detection of chemical contamination in the wells on the site will alter these plans it is too early to tell.[27]

As the Landsburg mine was being filled with hazardous wastes from the Duwamish Valley, the river itself was absorbing more than its fair share of the burden. Until after World War II, pollution generated by local industries had been relatively unregulated. The Rivers and Harbors Act of 1899 required upland disposal of solid waste and prohibited the unpermitted dumping of refuse into the river, but the law was focused on keeping shipping channels free of obstruction and did little to prevent other forms of pollution from fouling the nation's waters. Industries and cities with growing volumes of sewage to dispose of viewed the country's rivers, bays, and coastal areas as natural sinks for waste disposal—the dilution solution to pollution.[28]

Despite growing concerns about water quality, President Franklin D. Roosevelt vetoed the first water pollution control bill passed by Congress in 1938, choosing instead to extend technical and financial assistance to states to build sewage treatment plants and to encourage voluntary interstate agreements to reduce pollution. In 1948, Congress passed the Federal Water Pollution Control Act, which was signed by President Harry S Truman. The law provided both assistance and enforcement mechanisms to improve water quality. Despite the law's focus on direct threats to public health, a lack of political will and the absence of a clear system for setting standards and determining pollution allocations led to weak enforcement. The nation's water quality continued to decline.[29]

As industrial chemicals began to constitute an increasing volume of the waste dumped into public waters, they compounded and changed the nature of pollution. Instead of bacterial pollution from sewage, chemical pollution affected people, fish, and wildlife in new and insidious ways. Many of these effects—such as toxin-induced cancers—could take decades to manifest and were divorced from the public's day-to-day perception of declining environmental quality, but other effects were immediate and highly visible. One such example was the dramatic fire on the Cuyahoga River in Ohio in 1969—an event that is credited with spurring both the first Earth Day and the passage of the federal Clean Water Act in 1972. On the Duwamish River, public records show that although more than 150 riverfront industries were brought under pollution-restricting permits by the 1948 Water Pollution Control Act, the dumping of raw sewage and chemical waste into the river was still rampant. Several large oil spills also occurred on the river in the years preceding and following the passage of the Clean Water Act.[30]

The 1972 law has the stated goal of ensuring that all US waters are "fishable and swimmable." It amended and replaced the former Water Pollution Control Act, most notably by establishing clear nationwide standards for water quality and allowable pollution limits. The law was (and is still, despite amendments) widely considered far from perfect, but it introduced a radical new element that has proved to be one of the more effective—and controversial—enforcement tools in US environmental law.[31]

Section 505 of the Clean Water Act allows citizens to file lawsuits to force compliance when the state fails to do so, effectively deputizing the citizenry to enforce the law. This provision proved to be a key tool in controlling pollution in the Duwamish River, though not until a couple of decades after the Clean Water Act had been signed into law.[32]

FISH WARS

As new environmental laws were being developed on the federal level, fights once again broke out over fishing in and around Puget Sound, including on the Duwamish River. Despite increasingly tight state restrictions on fishing, including (and especially) by Native people, by the 1960s, sport fishers were complaining that those restrictions and the introduction of hatcheries had not saved the river's salmon runs. They began to blame pollution of the river for the decline of its fish.[33]

From 1958 to 1960, a series of fish kills left "rafts of dead salmon, trout, sculpin, and other fish pooling in the river's eddies," according to the Seattle historian Matthew Klingle. In 1959, the sport angler and conservationist Don

Johnson predicted "death for a tired old river." As runs continued to decline, the Washington State Sportsman's Council—a recreational fishing organization—sounded the alarm for the Duwamish River.[34]

In addition to pollution from riverside factories, the sports fishermen worried that a new government waste-management program would further deplete the salmon fishery. In 1958, King County voters approved the creation of Metro, a new regional sewer utility, to divert polluting sewage and runoff from Lake Washington. Pollution of the lake had become an aesthetic and public-health nuisance to the surrounding upscale neighborhoods. "All that sewage and wastewater had to go somewhere," wrote Klingle in 2007, "and that somewhere was the Duwamish River." The new agency opted to divert the waste into the Duwamish River, saving $18 million by avoiding having to build a long sewage pipeline leading out to the deep waters of Puget Sound.[35]

Metro was able to capitalize on fishing controversies to build support for this project. In 1962, three members of the Muckleshoot Tribe—successors to the upper Duwamish and Puyallup bands who secured a reservation on the Green River—were arrested for fishing below the Green River's Soos Creek Hatchery. As they awaited trial, another group of Muckleshoot fishermen took their place near the hatchery, which was sited on the grounds of a historical fishing village. These "fish-ins" were part of a coordinated campaign by the region's tribes to regain the fishing rights that had been guaranteed to them by treaty. Their actions enraged local sports fishers, who were determined to protect their privileged access to the fishery. Complaining about the "indigent bums" taking the river's salmon, they lobbied Governor Albert Rosellini to intervene.[36]

"Everyone knows that the treaty fishing rights are being interpreted far too broadly today," wrote the angler Jack Schwabland, insisting that "it never was contemplated . . . that Indians would mass gill-nets bank to bank on spawning streams and sell their catch commercially." (In fact, Native historians maintain that the region's Indigenous people may always have done exactly that.) The trial judge, however, acquitted the Muckleshoot fishermen, finding that the state had failed to prove the necessity of restricting Native treaty rights in order to save the fishery.[37]

The anger of the river's sports fishers after this decision gave the managers at Metro a perfect way to deflect blame for the decline of the fishery. In coordination with the state Fisheries Department, which had a long history of anti-Indian fishing policies, the agency launched a concerted public relations campaign to frame the threat to the river's salmon as the fault of overfishing by Native people. C. Carey Donworth, the chair of the Metro council, painted the Lake Washington sewer plan as a beneficial pollution-control project (which it was, for Lake Washington) and warned that salmon runs could "be ruined by uncontrolled Indian fishing on the eve of the end of the pollution threat." The Metro plan to reroute the region's sewer and stormwater lines to the Duwamish River steamed ahead, while the agency worked to direct the ire of sports fishermen and the public toward the threat of "uncontrolled" tribal fishing.[38]

The same year that the Muckleshoot won their day in court, a conflicting Washington State Supreme Court decision found that "modern" Indians had no privileged rights to fish in waters regulated by state law (or if they did, they could use only traditional gear available at the time of the treaties). Convinced that these conflicting decisions would ultimately be resolved in their favor, members of local tribes continued to fish and to pursue cases that would reaffirm their treaty rights.[39]

According to Stanley Moses, a son of one of these activists, the State Department of Fisheries intentionally frustrated the tribal fishermen's goals by repeatedly arresting them and confiscating their equipment, only to release them without pressing charges. One Duwamish tribal member who was cited and arrested several times for fishing on the Duwamish River was Charles "Manny" Oliver, a great-great-grandnephew of Chief Se'alth. Despite multiple fines and arrests, Oliver was never formally charged or given his day in court. A mass arrest of sixty Native fishermen on the Puyallup River in 1970, however, did eventually land the issue in court, leading to the landmark case known as the Boldt decision.[40]

Among the hundreds of people at a fish-in on the Puyallup River on September 9, 1970, was US Assistant Attorney General Stanley Pitkin. The tribes had been lobbying the federal government to sue Washington State in order to enforce their treaties, and they encouraged the attorney general's office to send a representative to the fish-in to witness the state's discrimination against Native fishermen. At the fish-in, Pitkin suddenly found himself in the middle of the action. Billy Frank Jr. of the Nisqually, one of the leaders of the fish-in—which took place on land owned by his family—described the day's events in a television interview years later. "We were fishing in the Puyallup River," when the police arrived with tear gas, he recalled. "They gassed us that day. They gassed all of us. They gassed the US Attorney, Stan Pitkin." Pitkin had indeed witnessed the state's approach to dealing with the tribe's treaty-protected fishing rights. Nine days later, Pitkin filed suit against the State of Washington on behalf of the US government.[41]

The case began in August 1973, supported by the testimony of Frank and dozens of other tribal members whose fishing rights had been restricted, along with scores of experts. Most of the state's commercial and sports fishing associations filed friend of the court briefs in support of Washington State, which was represented by Slade Gorton, the state attorney general and a future senator. In February 1974, the federal judge presiding over the case, Judge George H. Boldt, ruled in favor of the United States.[42]

Boldt upheld the supremacy of the treaties over state laws, delivering a "thunderous victory for the Tribes," according to the historians Walt Crowley and David Wilma. Even more stunning, he interpreted the treaty language that guaranteed the tribes' fishing rights "in common" with US citizens as meaning that the tribes had rights to an equal share—one-half—of all harvestable Puget Sound salmon. Since the tribes were catching only about 5 percent of the salmon harvested in Washington, the decision increased their share tenfold. Boldt ordered the state to immediately begin restricting non-Indian fishing in accordance with his ruling.[43]

For the tribes, it was a victory beyond all expectations. After watching their salmon runs decline for nearly a century, and having the state restrict their catch year after year, the tribes could finally regain access to their subsistence, commercial, and ceremonial salmon stocks.

Five years after the initial decision, Boldt's near-hero status among Puget Sound's Native fishing activists took a hit when he exempted a handful of the state's tribes from the fishing rights affirmed by his decision. The Duwamish, Steilacoom, Snohomish, Samish, and Snoqualmie Tribes had petitioned to be included in the list of treaty tribes benefiting from his order. Boldt referred the question to a magistrate, who determined that the petitioning tribes, who had not been part of the original case, did not qualify for fishing rights on the grounds that they had not maintained unbroken political cohesion since treaty times. The rejected tribes appealed the magistrate's findings to the judge. Boldt heard arguments in the appeal but then suffered a heart attack. He notified the circuit's chief justice that his health was failing and requested to be recused, but the tribes pressed Boldt to issue a decision. In 1979 Judge Boldt signed the last order of his career—reaffirming the magistrate's findings and excluding the Duwamish and other tribal petitioners from his 1974 fishing rights decision.[44]

Boldt reasoned that the tribes that had not been given reservations were not recognized by the federal government and did not qualify for treaty rights. In his decision, he wrote: "Only tribes recognized as Indian political bodies by the United States may possess and exercise the tribal fishing rights secured and protected by the treaties of the United States." The United States Court of Appeals for the Ninth Circuit later found this standard to be in error—federal recognition was not the issue—but they upheld the decision to deny the tribes fishing rights on the basis of the magistrate's report.[45]

The magistrate concluded that while individual tribal members could certainly trace their ancestry to a particular tribe, the original treaty tribes themselves no longer existed as political entities because of gaps in the historical records of the petitioning tribes' memberships and council organizations. Their treaty rights were instead invested in recognized "successor" tribes who had been granted and resided at reservations established by the treaties. In the case of the Duwamish, the US government today recognizes both the Suquamish Tribe—who occupy the original reservation to which the treaty required the Duwamish to move—and the Muckleshoot Tribe, which is named for the reservation later established along the Green River on Muckleshoot Prairie and consists largely of members of the upper Duwamish and Puyallup tribes.[46]

Manny Oliver ultimately enrolled in the Suquamish Tribe to secure the fishing rights he needed to support his family. He continued to exercise

Tribal canoes paddle up the industrialized Duwamish River during the Spirit Returns Paddle in 2002, signifying the Duwamish Tribe's return to their ancestral land. A Port of Seattle shipping terminal is in the background. Courtesy of Paul Joseph Brown.

those rights on the Duwamish River, purchasing a small fishing shack on the bank of the West Waterway (the canal flanking Harbor Island) with a group of friends. Despite the Boldt decision affirming their rights, they continued to be harassed by fisheries officials. In a 2013 interview with the *Muckleshoot Monthly*, Louie Ungaro, one of Oliver's fishing companions, recalled how fish and game officers had once hauled seventeen nets belonging to Native fishermen out of the water with a pickup truck. Judge Boldt issued a restraining order, but it was too late to save the nets. "We had a right to be there," Ungaro said, "and we got our nets back eventually, but they were all torn up by the rocks and were completely useless."[47]

Frustrated and disgusted, Oliver turned to other members of his family for help. He asked his sister, Cecile, to help with efforts by the Duwamish tribal government to protect Native fishing on the ancestral river. In 1973 Cecile Oliver (later Cecile Maxwell and then Cecile Hansen) began assisting the tribal chairman, Willard Bill, in his negotiations with state officials.

Two years later she was elected to succeed Bill as chair of the Duwamish Tribe—a seat she would hold for more than forty-five years.[48]

During the fish wars, Metro's rerouting of the city's stormwater and combined sewage overflows to the Duwamish River continued. Most of the raw sewage originally piped directly into the Duwamish River had been diverted through interceptor pipes to a new treatment plant built at West Point, north of Elliott Bay. There the sewage was screened to remove solids and discharged into deep water in Puget Sound. Lake Washington's sewage was diverted to the Green River, a short distance upstream of the Duwamish, where it was screened and treated with chlorine at a new plant in Renton. The West Point and Renton plants replaced roughly two dozen smaller sewage-treatment plants, handling all the sewage from Seattle and the Lake Washington suburbs.[49]

By the early 1980s, as much as 25 percent of the Duwamish River's flow during low water consisted of effluent from the Renton sewage plant. Stormwater and sewage overflows from an additional thirty-two square miles of the city now flowed to the Duwamish River instead of to Lake Washington. Water quality in the lake recovered at the expense of the river. The Duwamish Valley's residents and fishermen—increasingly comprising members of low-income, immigrant, and Native communities—still bore the brunt of pollution from the river's industrial factories and now had the added burden of treated and untreated effluent from distant neighborhoods.[50]

Finally, in 1987, the discharge pipe from the Renton plant was extended to Puget Sound, removing the treated sewage and its chemical by-products from the river. But at the end of the 1990s, the Duwamish River still received effluent from seven combined sewer overflows that together released an average of six hundred million gallons of untreated sewage and stormwater into the river each year. The outfalls were all on the last five miles of the river, contaminating South Park, Georgetown, a commercial tribal fishery made possible by the Boldt decision, and the city's former tide flats. "The sacrifice of the Duwamish," wrote Klingle in 2007, "hinged on displacing ecological costs onto the region's most marginal residents." Noting that a river cleanup effort had finally begun, he observed,

"Now Seattle must pay for the decision that the Duwamish was the best place to trash."[51]

During the Puget Sound fish wars, sports fishers and conservationists had been largely aligned in their efforts to "protect" Puget Sound's salmon by restricting Native fishing. After the Boldt decision, environmental and tribal advocates began to find common ground. The interests of the Duwamish Tribe and Seattle area environmentalists converged in 1974, when an electric transformer was dropped during a transfer between a loading dock and a barge on the river. An estimated 255 gallons of transformer oil contaminated with highly toxic PCBs (polychlorinated biphenyls, a class of industrial lubricants) sank to the bottom of the Duwamish. Initial cleanup efforts recovered about eighty gallons, but the rest dispersed throughout the side slip and into the main river channel. Two years after the spill, the Army Corps of Engineers dredged out ten million gallons of contaminated sludge, intending to dispose of the tainted sediment on Kellogg Island, a remnant of the river's original tide flats that sat directly across the straightened cut of the waterway from the spill site.[52]

In the intervening two years, the Corps of Engineers archaeologist David Munsell had discovered the Duwamish Tribe's ancient village site on the shoreline west of Kellogg Island. As plans to dispose of the contaminated mud were being developed, the Corps was excavating artifacts at the village and investigating the island as another possible site of archaeological significance. The right hand and the left hand of the agency were working at cross-purposes, complicating the Port of Seattle's efforts to fill in the last remaining river meander between Kellogg Island and the west bank of the river.

The discovery of the village also caught the attention of area environmentalists. The Seattle Audubon Society championed the cause of preserving Kellogg Island, which served as a sanctuary to sixty species of migratory birds. The society gained the support of Seattle City councilmembers John Miller and Randy Revelle and secured a unanimous vote by the council to conserve the island, only to lose the council's support when the Port of

Seattle threatened to sue the city. After voting for conservation status for the island, city councilwoman Phyllis Lamphere commented: "Even though a bird sanctuary is highly unusual in an industrial setting, the council considered such a use more desirable for Kellogg Island than a sludge depository." It took several years, more votes, and more studies before Kellogg Island gained permanent protected status.[53]

In the meantime, the Corps still needed a disposal site for its contaminated sludge. It was eventually buried in a pair of pits ten to twelve feet deep next to the river, on property owned by the Chiyoda Corporation. Disposing of the material on the eastern bank of the river may have protected the island and historic river bend from pollution, but it did not contain the toxic chemicals. Decades later, the buried PCBs were identified as an ongoing source of pollution to the Duwamish River.[54]

In 1966, a group of fishermen in New York's Hudson River Valley, angered by fish kills and ongoing pollution of their river, founded a watchdog group called the Hudson River Fishermen's Association. The organization was born of frustration—with the federal and state governments for not protecting their river and with New York City for actively polluting it with 1.5 billion gallons a day of raw sewage. In addition, dozens of factories dumped toxic waste into the river without consequence. The General Motors plant in suburban Tarrytown was known for dumping paint, turning the river a rainbow of colors that changed with the color of the cars coming off the production line.[55]

The US government's failure to enforce the 1948 Water Pollution Control Act left the fishermen with little recourse as the Hudson River's fish stocks plummeted. Recreational fishermen fly fishing with their children went home empty-handed. The livelihood of the river's commercial fishermen was threatened. When there was a good catch, the fish sometimes tasted of diesel. Third- and fourth-generation fishermen, some of them ex-marines and others with union experience from their day jobs in the trades, began to organize. In 1966, they called a meeting at the American Legion Hall in Crotonville, New York.[56]

One of the speakers was the former marine and *Sports Illustrated* writer Bob Boyle. He discussed a little-known provision of the Rivers and

Harbors Act of 1889, the law originally designed to prevent dumping of solid waste into the country's navigable waters and still in effect, which he thought they could use to stop the pollution. The law placed a bounty on illegal dumping for anyone who could document and prove it in court. Two years later, the Hudson River Fishermen's Association collected its first bounty, of $2,000, for documenting the dumping of diesel oil at a rail yard in Croton. With the bounty from subsequent cases, the fishermen eventually built a patrol boat, and in 1986 they formed a new organization dedicated to hunting down and prosecuting polluters on the Hudson River. By then, the 1972 Clean Water Act explicitly gave citizens the right to file lawsuits against polluters when the state failed to act. The new organization placed a fisherman named John Cronin at its head. Both the organization and Cronin were known as the Hudson Riverkeeper.[57]

The idea of the Riverkeeper caught on, and similar initiatives sprang up around the United States. Robert F. Kennedy Jr. signed on as the Hudson Riverkeeper's attorney, and in 1987, a local oysterman, Terry Backer, assumed the mantle of the Soundkeeper to clean up the Long Island Sound's fouled shorelines and oyster beds. In San Francisco, then the West Coast's largest port city, a Baykeeper took on polluters, using the funds collected from court-ordered fines and out-of-court settlements to fund environmental restoration and cleanup projects. Seattle joined these self-styled Robin Hoods of the waters in 1990, when the Puget Sound Alliance hired Ken Moser as the Puget Soundkeeper.[58]

Moser had been an adman, merchant seaman, and charter sailor. His personality was equal parts gregarious and irascible. He found a highly photogenic wooden boat, painted the words "Puget Soundkeeper" and a toll-free phone hotline number on its hull, and took to the waters of Puget Sound with his basset hound, Toby. He began to recruit kayakers and bird watchers to join a volunteer "environmental navy" and trained them to report pollution sources all over Puget Sound. Moser made national news when he sued the US Navy for polluting Sinclair Inlet, a small cove in Bremerton where nearly everyone worked at or catered to the naval shipyard. Robert F. Kennedy Jr. represented him, and they won their case.[59]

The Soundkeeper soon turned his sights to the Duwamish River and its wall of industry. In 1994, Moser and the local environmental attorney

Richard Smith discovered that unpermitted pollution was being discharged from two cement factories on Harbor Island, at the mouth of the Duwamish. They had allowed the raw materials for their manufacturing to pile up along their shoreline and fall into the waterway, polluting the estuary and threatening local fish and shellfish with arsenic-laden sand and gravel. The material also had pH levels high enough to kill herring, rockfish, and salmon as they swam through the chemical plumes caused by the spills. Soundkeeper sued, and in an out-of-court settlement, the company agreed to implement containment systems and a pollution prevention plan. The settlement also provided for automatic fines for any future permit violations.[60]

Moser was not the only one using the power of the Clean Water Act to go after Duwamish River polluters. Greg Wingard had turned his attention from old mining wastes in the Green River Valley to companies that were dumping toxic materials downriver. He founded an organization called Waste Action Project and set his sights on another cement company on the main channel of the Duwamish River. Holnam Cement sat on the land just south of Kellogg Island and the Duwamish Tribe's historic village, which had once been used by the Boeing Company and the US Postal Service to fly delivery planes out of Seattle. A hulking facility built in the 1960s, the factory was constructed of the concrete it produced. It was violating its waste discharge permit and polluting the river with the same chemicals and fish-killing debris as other cement plants a mile downriver. Wingard sued, and to settle its case, Holnam agreed to improve waste management practices at the plant and provide the Duwamish Tribe with funds for a habitat restoration project on Hamm Creek, a tributary of the Duwamish River.[61]

Ironically, none of these lawsuits could have been brought under the Clean Water Act if Washington State had been enforcing the requirements of the permits they issued: the act allows for citizen suits only when the state or federal agencies fail to enforce the law. Some in state government resented the citizen enforcers' intrusion into their domain; others applauded it. Regardless, it was poor enforcement of the Clean Water Act that opened the door for individual citizens to force the river's industries to stop their overlooked and illegal pollution of the river.

South Park's John Beal was an unlikely character to emerge as the Duwamish River's environmental champion in the closing decades of the twentieth century. A hard-drinking chain smoker with Coke-bottle glasses and yellowed teeth, he could often be found smoking a cigarette on a streamside rock while local schoolchildren planted saplings nearby or released juvenile salmon into the bubbling waters of Hamm Creek. But that was after the children had adopted Beal as their grandfatherly eco-savior and inspirational hero, after they had surrounded Beal while he told them about the regenerative power of nature and of their own power to heal the world around them. "This right here," Beal would say, using a stick to draw a circle in the dirt around their feet, "this is the environment. This is your environment. And what happens to it is up to you."

John Beal moved to South Park in 1976 with his wife and three young children. Born in Montana in 1950 and raised in Spokane, he never knew his father, who died of a heart attack a month after John was born. A learning disability and an inherently acerbic nature set Beal up for difficult teenage years. In 1967 he was expelled from high school, and according to his family, a local magistrate gave him the choice of going to jail or enlisting in the military. Despite his extreme near-sightedness and dyslexia, he shipped out to Vietnam as a marine rifleman to push back against the Tet Offensive in early 1968.[62]

After months of direct combat and multiple battlefield injuries, Beal wrote a letter home to his wife, Lana, a high school sweetheart whom he had married just before shipping out: "The doc seems to feel that I might need some mental care," he confided. "When I was hit, we were under mortar attack. He seems to think it might have done a little something to my mind." After recovering from his physical injuries, Beal was sent back into the field, joining a regiment with orders to level a jungle island with bombs and Agent Orange. There he earned the nickname of Johnny the Terror for his hand-to-hand combat, until he was captured, beaten, and locked in a cage as a prisoner of war. With the help of a local woman, he escaped after thirteen days in captivity and was sent back home eight months after being shipped out. He suffered from post-traumatic stress disorder for the rest of his life. At age twenty-nine, he suffered three heart attacks that may, ironically, have saved his life.[63]

In 1976, Beal's doctors diagnosed him with terminal heart disease, warning that he was likely to suffer another, potentially fatal, heart attack within months. "Get a hobby," they advised, hoping to channel his anxieties and prolong his life by a few months. Beal turned for solace to a deep ravine behind his house—the ravine where Dietrich Hamm's children had once fished. Here a murky tributary stream now flowed through discarded trash and blackberry brambles on its way to the Duwamish River. Thinking about the Vietnamese island of Go Noi where he had fought and denuded the riverbanks of their thick forest cover, Beal resolved to clean up the little pocket of creek in the time that remained to him.[64]

Beal began to drag washing machines, abandoned cars, construction waste, and everyday trash out of the stream. Digging out the blackberry choking the slopes at the bottom of the ravine, Beal read up on what kinds of plants he could bring in to replace the invasive bushes. He planted watercress, duckweed, and other native plants. Slowly, his private refuge became a rare pocket of native plant and wildlife habitat along the south end stream. He next turned his attention to the oily water that continued to flow through his hard-earned ecotopia.[65]

After a series of hit-and-miss efforts to filter out the oil, Beal dragged a hay bale down the hill and laid it across the narrow channel of water flowing between the saplings he had planted on the stream banks. It worked. In the following days, Beal visited Hamm Creek and watched as water with an oily sheen flowed along the creek upstream of the half-submerged hay bale and clear water flowed away from it. Beal continued reading, refined his hay-bale water filter with an oil-absorbing boom of his own invention, and began reaching out to scientists and government employees who might be able to help him restore his creek.[66]

As the creek began to thrive, Beal followed its flows upstream and down, removing trash and planting saplings as he went. He could only go so far, though: Hamm Creek traveled underground for much of its length, having been channeled into pipes and stormwater drains designed to keep the creek out of the way of businesses, streets, and homes as the Duwamish Valley transitioned from farmland to urban and industrial use.[67]

Beal continued to restore Hamm Creek for the next couple of decades, and his heart did not fail him. Having learned that insects are both a sign and a driver of ecosystem health, he harvested bugs from nearby streams and added them to the creek. He turned his backyard into a wildlife rehabilitation center, nursing injured beaver, otters, raptors, and owls brought to him by friends and neighbors before releasing them into the greenbelt around the creek. And he reached out to local businesses to enlist their help in preventing dumping and pollution. When they turned him away, he went around them or got creative. He met with state environmental officials to ask for help enforcing their anti-pollution laws and collected business cards from inspectors for the State Department of Ecology. Sometimes, according to Dan Cargill, a retired inspector, he used the cards to impersonate those inspectors as he tried to force recalcitrant creekside businesses to clean up their act.[68]

He also encouraged schoolchildren to raise and release baby salmon into the creek. When the salmon returned to spawn after their two to five years of living in the open ocean, he transported them by hand around the three-hundred-foot-long pipe that connected Hamm Creek to the Duwamish River at a forty-five-degree pitch, too steep for the fish to navigate by themselves. The freed salmon would then continue upstream to the creek's upper reaches. The children who raised the salmon in their classrooms also began to return each year. In turn, they inspired Beal to continue his work. But as Beal's vision grew, so did his need for help. Beal wanted to "daylight" Hamm Creek—to free it from its underground confinement and restore it to its original surface channel through South Park to the river.[69]

In the 1980s, John Beal approached the Duwamish Tribe to ask for help restoring salmon runs in the Duwamish River and its tributary stream in South Park. James Rasmussen, Quio-litza's great-great-grandson and a tribal council member, was particularly impressed with Beal's work and his passion for the river. The tribe appointed Rasmussen to be the council's liaison to Beal and his stream restoration efforts.[70]

Rasmussen's family had a long history in tribal leadership, dating to precontact days. The modern tribal council was organized to replace the traditional leadership structure that had been dismantled by colonization and displacement following the occupation of Duwamish lands. Rasmussen, descended from the high-born Duwamish woman Tupt-Aleut (Tyee Mary) and her Skagit husband, Kruss Kanum, was the third generation of his family to serve on the modern tribal council. His grandfather, Myron Overacker Jr., had served on it for thirty years since its inception in 1925. His mother, Ann Rasmussen, also served for thirty years. James had joined the council in the 1980s. "My name is James Rasmussen. That is a very good Indian name," he often joked when introducing himself to audiences who did not know his heritage or his history. "My family is very proud to say that we have been part of the leadership of this tribe for many generations."[71]

In 1984, the year he met John Beal, Rasmussen still lived on the Beacon Hill property homesteaded by his great-grandmother, Nellie (Quio-litza's daughter), and her husband, Myron Overacker Sr. Together, Rasmussen and Beal worked to focus attention on the river that had sustained Rasmussen's family for generations and the creek that Beal credited with saving his life. They organized a broad constituency of public and private interests to support the restoration of Hamm Creek and the larger Duwamish watershed. In 1990, in partnership with the City of Seattle and King County, they created the Green-Duwamish Watershed Alliance.[72]

With the tribe and local governments now providing assistance, Beal redoubled his efforts to save the creek. In 1995, King County agreed to purchase a reach of the creek where Beal had spent years working to remove trash and debris. On February 8, he wrote to Dietrich Hamm's son Lewis to let him know of the purchase. "Because of your father's dedication to preserving Hamm Creek," he wrote, "I thought you would be interested in knowing about this recent purchase and our future plans for the area." The project included a series of restored wetland ponds connected by fish ladders winding up the ravine where Beal had first discovered the creek. The project, named Point Rediscovery, was completed in 1998.[73]

But Beal's real passion was daylighting the creek. Across the highway from Point Rediscovery lay a farm once owned by Dietrich Hamm's neighbors, the Marra family, part of South Park's Italian farming community. The Marras had bought the land from Giuseppe Desimone and sold it to King County in the 1970s. Beal had traced the course of the creek water from the ravine at the newly restored Point Rediscovery into a storm drain, after which it briefly resurfaced across the highway before dipping underground at Marra Farm.[74]

When Wingard's Waste Action Project settled its lawsuit against the Duwamish River's Holnam Cement plant in 1996, Rasmussen and Beal used the funds to daylight the portion of the creek that ran underground at Marra Farm. The $36,000 settlement was donated to the Duwamish Tribe for the "Lost Fork" restoration, and the King Conservation District agreed to provide technical and permitting support. A five-hundred-foot reach of the creek was brought to the surface. Beal directed local volunteers in lining the newly daylighted channel with native wetland and riparian plants. Local farmers, mostly immigrant families growing foods native to their home countries, helped to steward the new restoration site, sometimes harvesting watercress and other plants for their own use. The completed project was dedicated in a ceremony in 2000. Hamm Creek became a symbol of urban waters brought back to life.[75]

I first met Beal when he was in the early stages of planning the Lost Fork daylighting project. I had been hired by the Puget Sound Alliance to organize volunteers for their "citizen navy," and in 1995, I took over from Ken Moser as the Puget Soundkeeper. Soon afterward, I received a hotline call about a fouling discharge into the Duwamish River. The call was from John Beal, who had spotted spills from what turned out to be an unlicensed ship-demolition company that had set up shop at the south terminal of the Duwamish Waterway. Together we documented their fly-by-night activities and their repeated dumping and spills and reported them to the State Department of Ecology. The state had become much more responsive to illegal pollution in the years since the Soundkeeper program began, and they shut the unlicensed operation down.

Beal later enlisted the City of Seattle's help in restoring Hamm Creek as well. The last thousand feet of the creek flowed through an underground pipe before spilling into the Duwamish River at one end of an untidy marina in Allentown, just past the southern boundary of South Park. Next to the marina sits Delta Marine, a manufacturer of luxury yachts, and farther south is a Seattle City Light substation. The latter consists of a collection of power transmission towers on land owned by the city, but it is located in an unincorporated pocket of King County, just south of the city limits. In 1997, Beal convinced Seattle to daylight the final connection between Hamm Creek and the Duwamish River.[76]

As salmon began to return to Hamm Creek, researchers at the Northwest Fisheries Science Center in Seattle's Montlake neighborhood were worrying about the effects of toxic chemicals on the region's surviving fish stocks. Salmon runs had plummeted following the industrialization of deltas like the Duwamish—where less than 2 percent of the estuary's native habitat survived the construction of the waterway and subsequent development—but scientists did not know what overall impact, if any, pollution was having on the Northwest's iconic salmon.[77]

Small habitat-restoration projects like Beal's were bringing back salmon by giving them a place to grow and to spawn, but the waters of Puget Sound's urban bays and rivers remained tainted with sewage and chemicals, depleting available oxygen and creating a toxic soup. No studies had been done of the possible effects of this pollution on salmon. After all, the thinking went, if the fish spent most of their lives in the open ocean far from Seattle and still returned home to places like the Duwamish, exposure to chemicals in the river obviously wasn't killing them. But the scientists at the fisheries center, a branch of the National Oceanic and Atmospheric Administration (NOAA), decided to investigate.

In 1989, Jim Meador snagged a coveted job as a postdoctoral fellow at the NOAA fisheries center, having just completed his PhD in aquatic toxicology at the University of Washington. He joined NOAA to work under the researcher Usha Varanasi, exploring the effects of chemicals like TBT (tributyl tin, a toxic organic tin compound), PAHs (polycyclic aromatic hydrocarbons, manufactured oils), and PCBs on fish in Puget Sound. Earlier work led by the science center's environmental conservation director, Don Malins, had documented high levels of these chemicals in Puget Sound fish, as well as associated abnormalities such as skin lesions and tumors. "The bottom line was that areas particularly near industrial activities and outflows had staggering amounts, in many cases, of toxic materials, PCBs, aromatic hydrocarbons, toxic metals, and so on," said Malins in a 2013 interview. "And associated with that was sickness in the fish." His team documented extensive evidence of pollution in both sediments and fish in the Duwamish estuary.[78]

More recently, NOAA scientists at another research station in Alaska had found that TBT was toxic to salmon being raised in fish hatchery pens treated with the compound. TBT was a popular antifouling agent, used to prevent barnacles and other mollusks from attaching to the hulls of ships—and the hatchery pens. Unfortunately, it was becoming clear that TBT was toxic to fish and that it migrated from the surfaces it was applied to into the marine environment. This turned out to be a particular problem at shipyards, where TBT-treated paint was intentionally removed during routine maintenance and often released into the surrounding water.

The dense concentration of shipyards in the Duwamish and its adjacent East and West Waterways created a corresponding TBT problem in the canals' bottom sediment. Partly as a result of NOAA's research, Harbor Island and the waterways flanking it were declared a Superfund site by the federal Environmental Protection Agency (EPA) in 1983. The area was among the first sites listed under the 1980 law that mandated identification and cleanup of the nation's worst toxic waste sites.[79]

Meador had conducted some of the nation's earliest TBT studies while serving in the navy in San Diego. When he arrived at the Northwest Fisheries Science Center in 1989, his job was to figure out how the chemical was affecting Puget Sound's Chinook salmon. Later, he was responsible for determining the environmental standards needed to ensure the endangered salmon's survival. Todd Shipyard on Harbor Island's West Waterway (the former Puget Sound Bridge and Dredge Company), which had been in operation since 1918, had released TBT, along with a host of other pollutants, into the channel. Meador set out to formulate a safe limit for TBT sediments that the company could be held to during cleanup efforts. But TBT proved to be a tricky chemical to pin down. The amount of TBT in the canal's bottom sediments was a poor predictor of how much of it was bioavailable, or could actually be taken up by fish. As a result, it was difficult to determine how much TBT could be in the sediments before harming the fish.[80]

Meador and his colleagues took a different approach: they measured the levels of TBT in the salmon and correlated them with signs of sickness.

But given the distances salmon travel, it was difficult to pin pollution in their bodies to any one source. If NOAA didn't know how much TBT in sediment would harm fish and couldn't prove that the harmful levels of TBT in the salmon themselves came from Todd Shipyard, they couldn't hold the company responsible for cleaning it up.[81]

In the end, the sediments contaminated by the shipyard's activities were cleaned up, but chemicals other than TBT were used to measure the project's success: arsenic, copper, zinc, PAHs, and PCBs. As the polluted mud was removed, the TBT was removed as well. As more was learned about the effects of the other chemicals on Puget Sound's fish, Meador and Varanasi redirected their attention to outbreaks of liver lesions in English sole in the Duwamish River that were linked to PAH exposure, and to the effects of PCBs on Duwamish River Chinook. Nearly a hundred years after the industrialization of the watershed began, there was no shortage of toxic pollutants for scientists to study in the Duwamish estuary.[82]

"Duwamish Cleanup Hailed as a Success," read the headline in the *Seattle Times* in May 1989. Even as NOAA was determining that the Duwamish River had an enormous pollution problem, local city and county officials were heralding the success of their efforts to stop that pollution at its source. "The flow of toxics—including fecal bacteria, chlorine, ammonia, copper, zinc and lead—that once made the river unsafe for marine life has declined sharply," the paper reported, pointing to Elliott Bay as the place where past pollution had settled. Metro, the county's wastewater agency, proposed spending $100,000 to start planning a cleanup of the bay, acknowledging that the effort would likely cost tens of millions of dollars. Yet cleanup of the Duwamish River, which supported a salmon fishery that was valued at more than $10 million a year, was reported as a fait accompli. "Metro said the improvements in water quality can be compared to success of [their] work in the 1960s halting the flow of treatment-plant effluent into Lake Washington," the *Times* reported. Mission accomplished![83]

Less than a year later, NOAA sued Metro and the City of Seattle. NOAA cited its own research showing that juvenile salmon in the Duwamish River carried elevated levels of PCBs, metals, oils, and other pollutants and alleged that Metro and Seattle had failed to control these chemicals in their stormwater and sanitary sewer systems. It was a bold move. "NOAA has traditionally been a scientific research agency that tracked environmental problems but did not correct them," the *Seattle Times* reported at the time. "Its decision to get involved in restoration puts a new player into the Superfund program to clean up toxic wastes." Indeed, the typical process is for EPA to identify a Superfund site and direct its cleanup before NOAA's involvement in restoration begins. On the Duwamish River, NOAA flipped the script and exercised its right to demand cleanup and restoration—plus control of ongoing pollution—in order to repair the damages to the Duwamish River's natural resources that their research had documented.[84]

A settlement reached the following year committed Seattle and King County to investing $24 million in sediment cleanup on the river, plus $5 million to preserve and restore habitat for salmon and another $2 million to control ongoing pollution of the river from their storm and sewer drains and the industries that used them. NOAA's attorney, Craig O'Connor, said that the total cost of cleaning up the river and bay would likely top $300 million, while the Washington State Department of Ecology's director, Christine Gregoire, said another fifty or more companies were also responsible for polluting the river and would be sued to help clean it up.[85]

River Revival

An Environmental and Cultural Renaissance

2000 TO THE PRESENT

FOR TWO FULL DAYS SINCE THE TERROR ATTACKS OF 9/11, THE SKIES over Boeing Field had been eerily silent. No planes flew in or out of the area that was normally inundated with the noise of jets taking off and landing at the King County Airport or passing over the river on their approach to Sea-Tac International Airport, about ten miles to the south. All flights nationwide had been grounded, and normal government services were effectively shut down. The country was in shock, and federal agencies were focused on ensuring that another attack was not imminent. Yet in the midst of this simultaneously frantic and frozen state of affairs, the gears of government ground on. On September 13, 2001, a notice appeared in the *Federal Register* announcing that the Lower Duwamish Waterway had been added to the National Priorities List, ranking as one of the nation's most hazardous waste sites. The Duwamish River was now on the Superfund list, and plans for a full river cleanup would begin immediately.[1]

Despite the events of 9/11, the Superfund listing was headline news in Seattle. The front page of the *Seattle Post-Intelligencer* bore the headline

Volunteers plant native shrubs along the Duwamish River in South Park during a "Duwamish Alive!" work party—a habitat-restoration event held at multiple sites along the river each spring and fall. Courtesy of Paul Joseph Brown.

"Duwamish Now on Cleanup List" on September 14, 2001, a century and a half, to the day, after the first white pioneers arrived to settle on the river. The Superfund site of 2001 was almost unidentifiable as the river where the Collins, Maple and Van Asselt families had first landed.

Dr. Usha Varanasi's findings of liver lesions in Duwamish sole, Jim Meador's research at NOAA on toxins in endangered Chinook salmon, and Don Malins's documentation of widespread fish contamination and disease in Puget Sound's urban estuaries had provided much of the evidence that led EPA to list the Duwamish River as a Superfund site.[2] By the time of the announcement, Meador's research was focused on PCBs— among the most toxic and ubiquitous pollutants in the Duwamish. Studies revealed that PCBs were present in all of the Duwamish River's resident fish and wildlife, despite a ban on their manufacture in the United States in 1979. These chemicals resist decomposition, so they persist in the environment for many decades. Alarmingly, and contrary to earlier assumptions, PCBs were even present in migratory salmon traveling through the Duwamish, including the small, silvery juveniles that passed through the estuary for just a few days or weeks on their journey to the sea.[3]

In 1999, EPA had directed a survey of bottom sediments along the five miles of the straightened Duwamish Waterway to determine the levels of contamination. PCBs were found in over 90 percent of the three hundred sediment samples—more than enough to warrant the Superfund listing. Local governments and businesses, however, wanted to keep the river off the Superfund list. They were worried about the stigma it would cast on the city, and some believed that they could clean the river up faster and better without EPA dogging their efforts.[4]

At the urging of state governor Gary Locke, EPA tried to negotiate a cleanup agreement with four of the largest "responsible parties"—those who caused and were liable for cleaning up the river's legacy of pollution. The four—Seattle and King County utilities, the Port of Seattle, and the Boeing Company—formed a partnership called the Lower Duwamish Waterway Group to formulate a plan that they hoped would keep EPA from designating the river as a Superfund site. EPA sought an agreement that would stipulate the same cleanup and restoration requirements as the Superfund law, without imposing all the federal red tape that would come with the listing. But Boeing refused to accept conditions related to paying for their historical damages to natural resources—the same part of the Superfund law that NOAA had used to require Seattle and King County to clean up and restore parts of the river years before.[5]

"The agencies wanted the right to claim payment for damage up to three years after completion of the cleanup, the same terms as allowed under the Superfund listing," wrote the *Seattle Times* in 2000. "But Boeing would not agree to those terms." Their refusal killed the deal. EPA listed the river on the Superfund National Priorities List, enacting all federal cleanup requirements and ensuring NOAA's ability to assess damages for the full period allowed by the law.[6]

A QUESTION OF SCIENCE

Meador's earlier research had found that PCB levels in Chinook salmon fry passing through the Duwamish estuary reflected the PCB loads in the sections of the river where they were caught. The salmon tended to hug one shore or the other and did not cross the channel. As a result, young salmon traveling down the east side of the Duwamish, where there were more factories and outfalls, had three to five times more PCBs in their bodies than salmon fry found on the west side of the channel. This discovery made it easier to figure out exactly where on the river the greatest harm was being caused and which entities were responsible.[7]

Even so, the task was far from simple. Wild Chinook salmon migrating from upriver streams had to be distinguished from salmon raised in the watershed's hatcheries, where it was discovered that they were being fed PCB-tainted fish pellets. In addition to poisoning the hatchery fish, the chemical-laden fish food undermined the scientists' efforts to measure the amount of PCBs picked up in the river. All of the hatchery salmon—marked as such by removal of their small adipose fin—had to be removed from the study to eliminate the influence of the fish pellets on the results. Once the data were limited to the wild fish, the pattern could be clearly seen.[8]

"It was really scary," says Meador. "The concentrations were really high in some of those fish." NOAA recommended that the areas with the highest PCB concentrations be cleaned up first, which they predicted would significantly lower PCB levels in Duwamish Chinook. This would help the Chinook to begin recovering before the entire river was cleaned up, which was expected to take decades. In response to Meador's findings, EPA selected seven "early action areas" with high levels of PCBs. Cleanup of these areas would be fast-tracked while the comprehensive studies, reviews, and engineering plans required for the full river cleanup were in progress.[9]

Thirty years after joining the team of scientists at NOAA, and nearly two decades after the river became a Superfund site, Meador is still at it. His most recent research shows that Chinook salmon migrating through the Duwamish and other contaminated estuaries of Puget Sound have a 45 percent lower survival and return rate than Chinook from clean areas. He hopes that new Chinook data will be collected now that cleanup of the early action areas is complete: the results would show how effective the early cleanups have been in reducing the PCB load in the bodies of the salmon.[10]

Meador removed himself from the increasingly charged technical meetings about how best to clean up the rest of the Duwamish River after EPA agreed to an approach proposed by Boeing and the other polluters that he saw as suspect. The actual scientific assessments were being done by the polluters directly, with EPA providing oversight. "It was interesting to see how they were spinning science in their direction," he mused in a 2019 interview. "I was skeptical that they would come up with a good plan." Agencies like NOAA reviewed the data collection plans in an advisory capacity but had no authority to approve or deny the methods used; that was up to EPA. Meador did not think the plans accounted for his finding that the feeding patterns of juvenile salmon were restricted to small areas. "They divided the river up into reaches, and then within those everything was averaged. It kinda diluted the whole thing." Despite his disagreement with EPA, however, Meador is glad to see that the Duwamish River cleanup is under way. "I'm optimistic that over time, it can be cleaner." After a pause, he added: "Hopefully, clean enough."[11]

Getting the complex science right is an integral challenge of every Superfund site, and especially of mega-sites like the Duwamish River. Scientific assessments, and the people who conduct them, are inherently biased, despite most scientists' best efforts to avoid bias in their research. Value judgments about what information is important guide the questions scientists ask and influence the conclusions that they come to. Judgments about the costs and time involved in a cleanup also come into play. Is it more important to be frugal and speedy, to get the cleanup done quickly, or to have a comprehensive understanding of all the long- and short-term ways in which pollution is affecting the fish, wildlife, and people in order to design the best cleanup plan possible?[12]

Not surprisingly, on the Duwamish River, there was a wide variety of opinions about how best to study and resolve the river's long legacy of pollution. The river flows through an urban landscape, surrounded by residential and industrial neighborhoods. It is used by recreational and subsistence fishermen from dozens of different nations and protected by tribal treaties with the US government, which effectively supersede all other applicable laws.

In addition, the cleanup raised complex scientific issues. The site contained a stew of forty-two different toxic chemicals in concentrations that exceeded safe levels for environmental or human health, or both. Were all these pollutants equally important? Some were found nearly everywhere in the river, while others were found in only a few spots. Which ones posed the greatest risks to fish, wildlife, and people? Could any of the chemicals be considered as markers for knowing whether the cleanup was successful? And most important, how clean was "clean enough"?

A BOTCHED JOB

In 2003, John Beal, the steward of Hamm Creek, was perched at the top of a rip-rap bank, an engineered slope of large boulders reinforcing the artificial shoreline under his feet. For days he had been watching the crane on a rust-colored barge anchored just offshore as its operator repeatedly lowered and lifted a metal bucket on a heavy chain. The base of the bucket consisted of hinged, interlocking teeth. Each time the bucket was lowered to the river bottom, the jagged teeth opened and bit deep into the soft mud. The bucket, heavy with contaminated sediment, was then raised above the water and swung around, opening its iron jaw to release a mound of slurry onto the barge. This project was the first EPA early action cleanup on the river.

Once or twice each day, a volunteer from nearby Georgetown or South Park would come by for a couple of hours to relieve Beal of his vigil. On a clipboard they recorded the time at which each bucket was lifted out of the river. In most cases they made notes about the volume of mud flowing over its top. Sometimes, if an obstruction prevented the teeth of the bucket from closing fully, it came up empty, the last of its slurry briefly seen as it escaped through the bottom. When that happened, the tainted mud was carried away by the current.

The sludge on the barge, and spilling into the river, represented decades of chemical pollution flowing from over 2,500 acres of industrial lands. In addition, the flow often contained untreated sewage, which gushed from a pair of concrete tunnels jutting out from the riverbank. Whenever the network of drainage pipes buried under the nearby factories filled with rain, a foul mix of industrial waste, sewage, and excess stormwater overflowed into the river. When the river was listed for cleanup, an average of more than three hundred million gallons of combined sewage and stormwater escaped each year from the pipes into the river. By 2003, system upgrades had reduced the volume to an average of sixty-five million gallons, overflowing roughly twenty times each year. Each overflow brought a new slug of PCBs, heavy metals, dioxins, and other chemicals to the river, just a few yards from where the accumulation of polluted mud was now being removed.[13]

The cleanup project was being conducted by King County. The outfall—called a combined sewer overflow (CSO)—was one of eleven within the five-mile Lower Duwamish Waterway Superfund site. Each of them carried polluted stormwater mixed with raw domestic sewage and partially treated industrial waste to the river during rainstorms. This particular overflow was named the Duwamish/Diagonal CSO, for its location at the intersection of the river and Diagonal Way South. EPA was officially supervising the work and had assessed the extent of contaminated sediment before the cleanup began. When the project was finished, the data would be used to determine whether the dredging had been successful.[14]

The project was also a test of EPA's promises of transparency and inclusion to local environmental groups, the Duwamish Tribe, and local residents. Whenever Beal saw the dredge bucket spill its toxic loads into the current, he made detailed notes. Whenever the spills increased in volume or frequency, Beal called the EPA manager supervising the operation. Beal and his volunteers—myself included—made dozens of calls over the sixty-seven days of dredging.[15]

John Beal had been a key player in forming a community advisory group to oversee the river cleanup. In the months leading up to the Superfund listing, the salmon habitat restoration group he had founded—the International Marine Association Protecting Aquatic Life (IMAPAL)—had partnered with the Duwamish Tribe and a team of environmental and social justice organizations to raise money to hire technical consultants and a coordinator for their oversight and advisory efforts. As the former director of the Puget Soundkeeper Alliance, I was asked to help manage the effort.

The coalition incorporated as the Duwamish River Cleanup Coalition (DRCC), and at Beal's urging, the neighborhood associations representing the residents of South Park and Georgetown soon joined it. DRCC's scientific consultants worked with Beal and others to examine the safety of King County's cleanup proposal. Several aspects of the plan caused the group concern, including the method of removing the most toxic mud. Instead of the specialized environmental dredgers in use elsewhere, King County's contractors used crude buckets that had not been designed to handle contaminated material. Indeed, when the dredging was finished, tests showed that while the targeted area was now clean, the surrounding river bottom—which had been relatively clean before the dredging began—was contaminated with PCBs spilled from the bucket.[16]

To reduce costs, King County had hired an unskilled operator who was inexperienced at working in environmentally hazardous conditions. As Beal's complaints to EPA accumulated, the agency ordered the county to slow down and exercise greater care. Yet these repeated demands failed to significantly improve the dredger's performance. Finally, the contractor agreed to switch to a state-of-the-art environmental dredge bucket. Relieved, we rushed to the riverbank to see the new bucket in action.

As the new bucket broke the surface of the water, it spewed mud from all sides. Beal stared in disbelief. Instead of an environmental dredge designed to seal off and prevent spills of contaminated mud, the new bucket had only a coarse metal screen. It was in fact a rock dredge, designed specifically to release all fine gravel, mud, and water—the polar opposite of a specialized environmental bucket. EPA issued a stop-work order, and the original open-top bucket was reinstated.

As the DRCC consulted with technical experts working on other Superfund sites around the country, they also became concerned that the work was being done before the full extent of pollution in the river had been studied. Because King County had been sued years earlier by EPA and NOAA for the environmental damages caused by outfalls like this one, the county had already planned to clean up the polluted area around this CSO. They were eager to get the job done before stricter cleanup standards could be imposed by the Superfund process. In the interest of speed, EPA agreed to permit the county to go forward under the rules of the previous settlement rather than follow all of the Superfund requirements.

Rushing to review King County's plans for the fast-tracked cleanup, DRCC was able to convince EPA to expand the cleanup zone to encompass a polluted sediment hot spot just upriver of the originally proposed boundary, so that the tainted mud there wouldn't drift into the cleaned-up river bottom. In addition, DRCC joined the City of Tacoma's mayor and council in opposing a proposed disposal site for the toxic mud from the Duwamish in Tacoma's Commencement Bay. Instead, EPA agreed that the material would be loaded onto railcars and transported by train to an eastern Washington landfill, permanently removing it from Puget Sound. However, the train carrying the first load of contaminated sediments derailed near the Columbia River, requiring another cleanup.[17]

At the end of the Duwamish/Diagonal CSO cleanup, EPA's test results confirmed the spread of contamination as a result of the dredging snafus. King County had to return to the river to remove the mud spilled by the dredge, at a cost of another $100,000. To the dismay of county officials, EPA refused to exclude the dredged area from future river cleanup plans, maintaining that the area might require additional cleanup to remove pollution still gushing from uncontrolled sewage and stormwater overflows into the Duwamish River.[18]

TURNING THE TIDE

Before the first bucket of contaminated mud was spilled at King County's sewage outfall, EPA was already drawing up plans for the next early action cleanup—a former asphalt company in South Park, about three miles upriver. In 1985, investigators Dan Cargill and Lee Dorrigan of the Washington State Department of Ecology had stepped onto a mud-filled lot owned by a company with the unintentionally ironic name of Malarkey Asphalt. Unpaved, pocked with pools of oil, and crisscrossed with exposed piping, the site had no visible pollution collection or treatment system. Ribbons of multihued runoff joined mounds of soft asphalt oozing toward the riverbank. Cargill and Dorrigan's report noted semiburied, dilapidated storage tanks scattered around the property. They collected samples showing that the adjoining beach contained high levels of lead, zinc, and arsenic.[19]

Once a seasonal floodplain prized by early settlers for its fertility, this reach of the Duwamish River was dredged in 1918 as part of the waterway project to straighten its meanders and deepen its draft. Here the canal made a thirty-degree eastward turn before continuing its straight line south. The adjacent floodplain and a neighboring river oxbow were filled with dredged mud, rocks, and debris to create new land. A couple of homes were built here as the neighborhood of South Park expanded. In 1937, the Duwamish Manufacturing Company bought the property, removed the houses, and began manufacturing asphalt roof shingles. In time, the company changed hands and became Malarkey Asphalt Co., named for its principal owner, Michael Malarkey. It ceased operations in 1993 and later sold the property to the Port of Seattle.[20]

I had passed the Malarkey Asphalt property dozens of times while patrolling for pollution as the Puget Soundkeeper in the 1990s, and later while leading environmental education tours on the river for the DRCC. The most dramatic feature of that part of the riverfront at that time wasn't the asphalt plant but the fifty-foot-high tanks owned by the Basin Oil Company, across the street from the shuttered plant. In the years immediately following the Duwamish River's listing as a Superfund site, John Beal frequently called the Puget Soundkeeper's pollution hotline to report mysterious burning in the tanks at night, or spills of oil running into nearby street drains. Historical pollution on the Malarkey site—now known as the Port of Seattle's Terminal 117—had triggered an EPA cleanup order, which the port completed in 1999. After that cleanup, the site had become a quiet parcel of land leased out for truck parking and storage.

In 2003, despite the previous cleanup, the old Malarkey riverbank made EPA's list of hot spots within the Superfund site that posed an immediate threat to the environment and human health. Planning for a new early action cleanup began. In the world of Superfund cleanups, however, *early* is a relative term, and after seeing the results of King County's cleanup at its sewer outfall, Terminal 117's South Park neighbors did not want to risk sacrificing quality for speed. Residents took a special interest in overseeing the investigation and cleanup of the waterfront property. Any spillage here could wind up on local public beaches or at the shoreline park just a short distance downriver.[21]

Guy Crow owned a marina next door to Terminal 117. As EPA worked with the port on a new cleanup plan, Beal and I often sat with Guy to talk about his recollections of Malarkey. Over two decades, he had seen company staff dump asphalt sludge onto the riverbank, bury drums of oily waste on the factory grounds, and pour hundreds of gallons of waste oil into an old railcar they used as a makeshift holding tank to heat the oil before pumping it through a series of pipes to machinery in need of lubrication. It was the railcar that had been the focus of EPA's cleanup in 1999: the tank was unsealed, and the oil seeped out into the surrounding soil and groundwater. In the 1990s, the EPA investigation was confined to the footprint of the railcar, which had been removed. The thick blackberry hedge between the railcar and the riverbank was not included.

Seeping out from under the hedge, and visible from the water, I had seen the lava-like flows of asphalt that would ooze sludge if pressed gently with the toe of a boot, along with the random barrels toppling out of the bushes at the top of the bank. The buried drums, however, I had heard about only as rumors from Crow and one or two other longtime neighbors.

If the previous upland cleanup had failed to remove the visible drums and spilled asphalt, what else might still be buried unseen on the site? Residents' worries were exacerbated by the lack of soil sampling under much of the property's thinly paved parking lot, and even more by a single sample taken at the edge of the pavement near the bank, which showed startlingly high levels of PCBs: 1,000 parts per million (ppm), a thousand times higher than the state standard for a residential area. When EPA released its draft early action plan in 2005, it proposed to remove the contaminated riverbank up to the edge of the parking lot but no farther. Any PCBs that might lie buried inland of the riverbank would be left alone.[22]

The DRCC knew that asking for more sampling would add time to the cleanup, but the high levels of PCBs at the site could pose a long-term risk to both the environment and human health. On April 7, 2005, the coalition wrote a response to EPA's cleanup plan. "Unfortunately, the site is inadequately—indeed, incompletely—characterized. . . . DRCC is extremely concerned that EPA and Ecology are sacrificing quality for speed in their first cleanup proposal for the Duwamish River, and we are unable to support the proposed action." In addition, the coalition noted, "The public cannot provide informed comment on the adequacy of an incomplete plan." The DRCC asked EPA and the port to conduct more testing under the parking lot where Crow claimed that barrels and transformers had been buried. Separately, it also requested that the drums and sludge along the shoreline be removed immediately to prevent any remaining oil from seeping into the river.[23]

EPA and the Port of Seattle insisted that there was no cause for concern. The agency's published reports stated that the previous cleanup on the property had reduced all PCBs in the soils to 25 ppm or less, and that at an industrial site, 25 ppm was a level low enough to protect people's health. They viewed the 1,000 ppm sample at the top of the bank as an anomaly. Fortunately, though EPA had ordered the cleanup, it was also required to satisfy all state laws. Washington State required PCBs to be cleaned up to levels of 10 ppm or lower on industrial properties and to 1 ppm or lower on any property within one hundred feet of homes.[24]

To reassure the community, the Port of Seattle sent a team of technicians to Terminal 117 to take more samples near the hot spot at the top of the bank. The technicians collected samples by punching small holes through the four-inch thick asphalt covering the parking lot. A short distance inland from the bank the samples revealed PCB concentrations of 1,400 ppm. Levels were increasing rather than decreasing as crews worked their way toward the property boundary, the public right-of-way, and the houses across the street.[25]

Guy Crow was watching a sampling crew in the parking lot at Terminal 117 in March 2006 when they suddenly jumped back from the hole they had punched through the asphalt. They turned and rushed off the property, outside the fenced barrier the port had erected a few weeks before. When they returned, they were wearing protective suits. The samples they took, just a few feet from the fenceline and within one hundred feet of the closest homes, showed levels of 9,400 ppm of PCBs—nearly ten times higher than the levels at the top of the riverbank. The *Seattle Post-Intelligencer* quoted the port spokesman David Schaefer: "Everybody was surprised. . . . We may need to dig out more material."[26]

In the meantime, Greg Wingard, now executive director of the non-profit group Waste Action Project, planned a site visit to count the number of fifty-five-gallon drums sitting along the riverbank below the old Malarkey plant. Volunteers, myself included, had noted on several trips up the river that at least one—a rusty drum clearly visible from the water—appeared to be dripping oil. Accompanying Wingard to the site, I brought sterile bottles for collecting samples of leaking oil. If EPA and the port wouldn't test the drums, any lying below the high-tide mark on the public shoreline were fair game for us to sample.

We arrived by kayak, pulling up on the shoreline at low tide. Prodding at the blackberry bushes at the top of the slope, Wingard counted a couple of dozen drums, most in a state of advanced corrosion. One drum was buried deep in the beach at our feet, with only the top couple of inches visible above the sand. The barrel we had noticed on previous boat trips was perched halfway up the bank, supported by a collection of rotting timbers jutting out over the water—remnants of an old retaining wall.

A visible sheen dripped from the exposed barrel. Wingard pulled on safety gloves and collected a sample in a small glass vial. A nearby lab, which had offered to run a chemical profile of any samples we collected, delivered the results a few days later. The sample tested positive for PCBs. We forwarded the results to EPA and the Port of Seattle, requesting immediate removal of the barrel and any others on the riverbank that might still be leaking PCB-laden oil into the river.

Following the new discovery of high levels of PCBs near the Terminal 117 fenceline, EPA ordered the City and Port of Seattle to extend their sampling beyond the property line, into the unpaved city streets and residential areas nearby. PCBs were found in dirt in the streets and parking strips. Residents' private yards and gardens showed levels ranging from 20 to 93 ppm. A handful of homes were also tested and showed low levels of PCBs inside.[27]

In light of all the new data, EPA developed a revised cleanup plan that would include the upland property and neighboring streets. The City of Seattle's power utility—Seattle City Light—was added as a responsible party to help pay for cleanup because the original source of the oil used by Malarkey was waste oil from the utility's power transformers.[28]

Based on the new sampling, EPA released a second plan to address the upland property in order to cover the entire area of contaminated soil, sediment, and groundwater. The front yards of two homes across the street were removed and replaced with clean soil in an emergency action. In addition to cleaning up the polluted beach, deepwater sediments off-shore, and the riverbank, the new plan would remove the top layer of PCB contamination throughout the uplands (plus a few deep bores where PCB levels were highest). It would also replace all the dirt roads with clean pavement on the streets surrounding the Malarkey plant. It specified that PCB levels on the Malarkey property not exceed 10 ppm at the surface and 25 ppm at depth. The plan would restrict all future land uses to industrial and require that the entire property be paved and fenced in perpetuity to protect the public from any health risks that remained.[29]

The South Park community was stunned. After discovering the extent of toxic PCBs at the site, EPA intended to leave ten times more pollution near their homes than state law allowed. In doing so, EPA and the port were limiting how the community could use their waterfront, no matter who might ultimately own the narrow strip of land. The decision would ensure that no one could ever buy the land from the port without also taking on the liability that would come with the PCBs left behind.

The DRCC again responded to EPA's plan: "The Port, as a public agency, must take public concerns into account before deciding that this property will forever be highly restricted industrial property," they wrote. "This predetermined land use does not maximize the potential for re-development opportunities that may better serve the South Park community."[30]

Residents also complained to their city councilors. Technically, Terminal 117 was in unincorporated King County—a remnant of zoning years before that had left a narrow strip along South Park's riverfront surrounded by, but outside, the city limits. The city had planned for years to annex the "Sliver by the River." But EPA's decision would limit the city's options for future land uses along the river, despite spending millions of taxpayer dollars to clean it up.[31]

At EPA's public hearing on the proposed plan in May 2006, residents showed up in force and spoke out against the plan, and the Seattle City Council presented their objections: "The City Council is currently reviewing the area including the T-117 for annexation into the City of Seattle." Citing future commercial and recreational possibilities, they added: "We are interested in ensuring the land is clean enough to be used for any zoning designation that the city may determine appropriate."[32]

In a last-ditch effort to change the plan, the DRCC held a series of one-on-one meetings with members of the Seattle Port Commission. In addition to raising environmental justice issues—challenging them to consider whether they would support such a decision near the homes of Seattle's wealthier residents—coalition members emphasized the financial implications of the permanent liability the commission would incur by leaving PCBs buried on their property. Until then the port had been largely unresponsive to environmental concerns. However, Seattle's newspapers had been sympathetic to the low-income South Park neighbors' plight. On the day of the Port of Seattle Commission meeting, where the port was expected to sign its agreement to carry out EPA's cleanup order, South Park residents carpooled en masse to the commission's meeting seven miles south, at Sea-Tac International Airport.

The large, usually empty, meeting room was packed. Agency staff and reporters lined the room. Residents lined up to testify. One by one, they decried the port's plans to leave toxic chemicals buried in their neighborhood. One mother broke down crying as she described walking and playing with her young children on the streets around the Malarkey property, where they had been unknowingly exposed to PCBs for years. Others accused the port of environmental racism for its decision to leave this largely Latino community burdened with toxic pollution.[33]

Newly elected port commissioner John Creighton voiced his concern that failing to fully clean up the site would expose the port to more liability and added, "I do think we need to earn back the trust of the community." Each commissioner speaking after him then voted to clean up the site more than EPA's plan required. Paradoxically, the Port of Seattle commissioners were unanimously rejecting EPA's cleanup plan on the grounds that it did not require them to do enough—even though the EPA plan had been largely written and designed by the port's own staff.[34]

Dan Cargill, the state inspector who had first investigated the site in 1985, was reviewing the T-117 cleanup plan on behalf of Washington State and had watched as EPA approved a plan that failed to meet state standards. When the port rejected the cleanup order, "We were surprised and overjoyed," he later recalled. "We had been trying to get EPA and the Port to understand that the levels they selected would be an ongoing source of recontamination" to the Lower Duwamish Waterway. His frustration was compounded, he said, because he was one of the few people who had known just how badly polluted the site was when Malarkey was still operating, and had shared his observations with EPA. "We knew it was far more contaminated than the City, Port, or EPA thought it to be. Our concerns and comments were ignored, rebuffed, or dismissed at every turn."[35]

The Port and EPA managers responsible for the rejected plan were removed from the project. The new managers—Roy Kuroiwa for the Port of Seattle and Piper Peterson for EPA—welcomed the involvement of the community. The Port of Seattle commissioners held their next monthly meeting in South Park—the first they had ever held in a neighborhood affected by their activities. The top item on the agenda was the neighborhood's vision for the port's property. Over a hundred South Park residents attended and unanimously endorsed a plan that would turn the entire property into a salmon habitat restoration project, with public viewing platforms and a small boat ramp.

Given the new positions taken by the Port and the City of Seattle—the parties who would pay for most of the cleanup—EPA went back to the drawing board. In June 2010 it issued a new cleanup order requiring full removal of all PCBs above a level of 1 ppm. Later that year, the port began planning a full site cleanup to haul out the contaminated soil and create off-channel river meanders that would eventually support the Duwamish River's migrating salmon.

In August 2013, port contractors hit something hard as they dug deep into the contaminated dirt at Terminal 117. More than forty buried drums of toxic waste, never detected by the port's sampling probes, were discovered buried under the site of the old asphalt plant and removed.[36]

BOEING'S PLANT 2

In the years it took EPA and the Port to finalize a cleanup plan for Terminal 117 that the community would support, the City of Seattle had quietly planned and, in 2012, finished a cleanup of Slip 4, a remnant river meander in Georgetown, across the channel from South Park. Slip 4 borders Boeing's Duwamish River property and receives drainage from Boeing as well as City of Seattle and King County CSOs and storm drains. The PCB levels found at the slip, along with dioxins and heavy metals, landed it on EPA's early action list. It was here that Jim Meador and his NOAA researchers had collected their most highly contaminated salmon.

Georgetown residents had been watching South Park's fight over the cleanup of T-117, and the city didn't want a repeat performance in the historically scrappy neighborhood with ties to organized labor and City Hall. To avoid it, they consulted with residents at each step in the process. The result was a cleanup plan for the slip that everyone could agree to: in fact, the standards set at Slip 4 became a model for future cleanup actions, and a smooth approval process with community support facilitated a fast cleanup. Within a few years of EPA's proposed plan for Slip 4, river otters were seen recolonizing the previously contaminated riverbank. The remaining challenge was ensuring that the slip wouldn't be polluted again by PCBs from an unidentified source.[37]

One of the drains into Slip 4 ran from the north end of Boeing Field and under the city's historic Georgetown Steam Plant, which had lost its water intake when the river was straightened. The airfield had been used by Boeing since the 1930s, when it moved to the land bought for one dollar from the South Park farmer Giuseppe Desimone. Over the years, Boeing had used caulk containing PCBs to seal everything from blocks of tarmac to windows in its warehouses and business offices. After an extensive search, investigators discovered that the PCBs in the caulk were leaching out and into stormwater runoff—right into the flume running from the steam plant into Slip 4.[38]

The pipe belonged to the city, and Boeing initially denied any responsibility. Seattle accused Boeing of illegally using its drain for their polluted runoff. "There have been over twenty lines attached to our ditch that came from the Boeing Company," complained the city utility manager, Martin Baker. Eventually, the PCBs were traced to Boeing. The company dug out all the old caulk from its airfield, but the chemical had seeped into the surrounding concrete and continued to contaminate Boeing's stormwater. Finally, EPA required the company to install a PCB treatment system to cleanse the water in its drain before it hit the river.[39]

Boeing's new treatment system solved the threat to the newly cleaned Slip 4, but the company had a much larger problem stretching southward. EPA had flagged more than a mile of riverfront fronting Boeing's historic Plant 2 as an early action cleanup area. It was the largest and most highly contaminated reach of the Duwamish River, with toxic chemicals found buried at depths of more than four feet below the river bottom and dozens of sources of ongoing pollution scattered throughout the property.[40]

Even before the Duwamish River was listed as a Superfund site, cleanup plans were under way for Plant 2. Decades of airplane manufacturing, particularly before the environmental laws of the 1970s, left high levels of toxic material throughout the plant. During World War II, when the plant was producing as many as seventeen B-17 bombers a day, pollution control had not been the priority, and employees described waste-management practices continuing into the 1970s that consisted of sweeping the debris from the factory floor into the river at the end of each shift.[41]

The EPA had ordered Boeing to clean up Plant 2 and its adjacent river-bank in 1994, under the requirements of the Resource Conservation and Recovery Act (RCRA). RCRA requires cleanup when hazardous wastes have been improperly released or mishandled. This was Boeing's preference: the company had fought to keep Plant 2 off EPA's Superfund site list. Even so, Boeing challenged and resisted EPA's oversight at nearly every step.[42]

The company had a reputation for dragging its feet and attempting to intimidate regulators at dozens of its toxic-waste sites around the country. Its representatives fought with the City of Seattle over who was responsible for polluting Slip 4, and now they fought with EPA about the standards for the Plant 2 cleanup. A key point of contention was whether people should be able to fish in the Duwamish. "I think we need to set reasonable expectations for cleanup in industrial areas," said Boeing's Steve Tochko in 2009, scoffing at the idea of fishing and industry coexisting on the river. Shawn Blocker, the EPA cleanup manager for the site, saw things differently. "Boeing doesn't feel that stretch of the river could ever be restored to where you could harvest these kinds of fish. We disagree with them."[43]

In 2006, a shift in attitude at the company began to change this long-standing dynamic. Boeing's legal battles against EPA and states had become so expensive that they decided to commit funds to clean up their pollution in order to rid themselves of legal bills. Boeing's top executives allocated half a billion dollars to clean up dozens of contaminated sites in the Pacific Northwest. They proposed cleaning up Boeing Plant 2 by removing some contaminated sediments and then building engineered caps of sand and gravel to isolate the material left behind. Cost estimates for this approach were between $50 million and $60 million for Boeing's mile of riverfront.[44]

Shawn Blocker wanted more, but his authority was limited. Technically, Boeing's cleanup proposal would meet the requirements of the law. EPA could, however, require the company to prove that it had the financial resources to maintain and monitor the caps containing the buried waste. Such assurances were usually required for a period of thirty years, but on occasion EPA stipulated longer periods. At the Iron Mountain Mine in California, EPA had required the property owners to show they had the ability to monitor and contain the remaining toxic chemicals for three hundred years. Blocker approved Boeing's plan but told the company he would require a one-hundred-year period of financial assurance for the toxic material they left in the Duwamish River.[45]

Like an insurance policy, such assurances cost money. To meet EPA's requirements for a financial guarantee, Boeing needed to provide a letter of credit from a bank. The bank's price to "insure" Boeing's caps was $49 million—nearly as much as the cleanup itself. Boeing did the math. Overall, it would be cheaper to remove the remaining contamination than to pay the cost of not doing so. Boeing decided to remove all PCB contamination above 2 parts per billion (ppb)—the standard set by Slip 4—thus avoiding the need to purchase any insurance for the remaining contamination; there was none.[46]

EPA had not ordered a full cleanup of all of the contamination Boeing had trickled and dumped into the river over its century of airplane production. That the company decided to remove it was purely a business decision, designed to avoid future costs. Two of the river's polluters—one public agency and one private industry—had now decided to clean up more of their historic contamination than EPA had required. What would the standard be for the rest of the river? Perhaps more important, who would decide?

A VISION OF HEALTH

Trace de Garmo walked along the damp paths of trampled ground to rouse his fellow homeless campers from their tents on a chilly spring morning in May 2012. Hot coffee and doughnuts were laid out on tables next to a makeshift pantry that was stocked with donated cans of food, bread, and paper towels. A team of staff and interns from the Duwamish River Cleanup Coalition waited to greet the tent city residents with clipboards, maps, and a colorful stack of "Duwamish dollars"—credits redeemable at nearby convenience stores and cafes.

The collection of tents tucked into a shallow depression between the First Avenue Bridge and the Duwamish River was known as Nickelsville—recalling the Hooverville that existed on the tide flats in the 1930s and playing on the name of Seattle's former mayor Greg Nickels. The residents called themselves Nickelodeans. The encampment had been moving around Seattle since 2008. They had been evicted from this same piece of property by Nickels several years earlier, and they and their supporters held Nickels responsible for making the city increasingly inhospitable to homeless residents. Under the new mayor, Michael McGinn, the city pretty much left them alone.

Anna Schulman, a student at Antioch University, poured cups of coffee and explained the reason for the coalition's visit. The DRCC was interested in learning about any health concerns Nickelsville residents had while they were camped near the river. Staff members Paulina Lopez, a South Park resident from Ecuador, and Alberto Rodriguez, originally from Honduras, recorded the residents' answers on survey forms and maps. The information would be used to help the residents decide how to use a small community health grant to improve conditions in their encampment.

The Nickelsville effort was part of the DRCC's Duwamish Valley Community Health Project. The Terminal 117 cleanup campaign had shown that the future of the Duwamish waterfront could be very different from its present state. EPA was now making plans for the rest of the river, and DRCC was seeking input from the community about their vision for the waterfront.

The Superfund site, which extended along both shores of a five-mile length of the river, affected the activities of residents, businesses, workers, fishermen, tribes, and a growing number of recreational visitors. What did these people want the river to look like? Collecting input from interviews, workshops, surveys and focus groups, along with information from twenty previously published river basin plans that addressed topics from green space to affordable housing, the coalition published its final report on the community's vision for the Duwamish Valley in 2009.[47]

The vision was wide-ranging: cleaner and greener neighborhoods, a revitalized business and industrial base, improved transportation networks, and strategies to ensure that Duwamish Valley residents could afford to remain in their neighborhoods after the cleanup. But the overarching message was simple: people wanted the river cleanup to catalyze holistic and wide-ranging improvements in individual, environmental, and economic health.[48]

The coalition knew that reducing toxic pollution in the river bottom would not be enough to achieve this goal. The valley was riddled with pollution sources on land and in the air; low-income residents had poor access to health care; and if cleaning up the Duwamish attracted developers who gentrified the waterfront, residents and businesses alike could find themselves displaced. Community members had a deep fear that after fighting for a cleaner and greener Duwamish Valley, they wouldn't be the ones who ultimately benefited.

The report articulated particular concerns of and for the most disadvantaged members of the river's community. By the early 2000s, generations of new immigrants from Europe, Latin America, Asia, and Africa had made the Duwamish Valley's zip code one of the nation's most diverse. Children in the local schools spoke more than thirty languages. Most were from low-income and working-class families. Residents of homeless encampments fished in the river alongside sports anglers and tribal members. The report committed to a principle of "acting with compassion for neighbors and others in need."[49]

The DRCC had secured a grant from EPA to help implement some of the community's ideas. They focused on improvements that were unlikely to be addressed by the Superfund cleanup itself, including health improvements to the Nickelsville encampment—to be determined by the camp's own residents. Other projects included tree planting, food banks, and support for youth and multiethnic programming at the historic Marra Farm, now a city-owned community garden used by many of the valley's immigrant families. But the EPA grant would not be sufficient to address the community's more challenging needs.[50]

In March 2013, Bellamy Pailthorp of the public radio station KPLU reported on "more illness, shorter lifespans in Duwamish Valley." A Seattle health-advocacy organization, Just Health Action, in cooperation with DRCC, had just released a report on local environmental and public health. The report, funded by an EPA environmental justice grant, was the first to compile localized health data. It compared disease rates in different parts of the city as well as indicators of environmental and social health like income and access to parks. The study examined "the cumulative health impacts of exposure to pollution and other factors," Pailthorp explained, particularly those "known to make people more vulnerable to illness, such as poverty and stress."[51]

The study detailed the presence of more toxic waste sites in the Duwamish Valley than elsewhere in the city and higher levels of diesel pollution in the air. It found that asthma hospitalization rates here were also among the city's highest, validating the concerns of residents who had long suspected that their children suffered disproportionately from the disease. The most striking finding was that the average life span of people who lived in South Park and Georgetown was eight years shorter than the city average—and a full thirteen years shorter than that of residents of the study's healthiest neighborhood, Laurelhurst, which was also one of its wealthiest and whitest.[52]

In a television interview on *KOMO News*, the lead researcher and author of the report, Linn Gould, commented: "This study shows that Duwamish Valley residents are disproportionately and unfairly burdened by multiple stressors outside of their control. . . . Decision makers should take action to resolve these inequities." James Apa, a spokesman for the city and county health department, agreed that "place matters" and stated that the department's job was to develop solutions to eliminate these disparities.[53]

The DRCC urged the city to create a community health task force with a dedicated fund. It also recommended scrutinizing the Duwamish River cleanup plan in light of the new data on the valley's pervasive health disparities. Linn Gould and Dr. Bill Daniell of the University of Washington's School of Public Health stepped forward to help. Daniell secured a grant from the Robert Wood Johnson Foundation and Pew Charitable Trusts to conduct an independent study of the health impacts of EPA's cleanup plan for the river. Time was of the essence: EPA had released its draft cleanup proposal for the full cleanup of the Duwamish Superfund site the month before.

Daniell's team had started doing background research for their health impact assessment (HIA) of the cleanup plan several months before it was formally released for public review. As part of that team, and in my capacity as DRCC's health program manager, I was working with Jonathan Childers, a doctoral student in public health, to examine the implications for local residents. Other team members were researching implications for the tribes, subsistence fishing families, and workers in the Duwamish Valley.[54]

As EPA was getting ready to make its cleanup plan public, King County published its own assessment of expected impacts. As part of the Lower Duwamish Waterway Group—the polluters who had to pay for the cost of cleanup—the county argued that less removal of contaminated sediments would be better for residents' health and well-being. The county asserted that extensive cleanup would increase truck traffic through the riverfront neighborhoods (which were already trying to reduce freight traffic on their streets), and levels of diesel air pollution—already disproportionately high in the Duwamish Valley. These predictions, packaged as an "equity review" of the project, struck directly at the heart of riverfront neighbors' fears.[55]

Not everyone in county government agreed with the reviewers' approach or conclusions. The Seattle-King County Department of Public Health criticized the report for "lacking an analytical review." They also cited the county's Wastewater Division for conducting the analysis in isolation, suggesting that a single person had done the analysis in a single sitting, and for failing to involve members of the affected communities—a violation of the core principles for all of the county's equity impact reviews. Three pages of specific technical and procedural criticisms were given to the county's review team by its public health department after the report was released.[56]

Hurrying to complete their work, the HIA researchers learned from King County advisers that trains, not trucks, would be the primary means for transporting sediments dredged from the river. In addition, they learned that EPA rules did not allow diesel fuels to be used in those trains. These facts directly contradicted the neighborhood impacts cited in King County's report. The HIA attempted to set the record straight, but in some respects it was too little, too late.

Representatives of King County and the Lower Duwamish Waterway Group had been publicly airing their claims about truck and air pollution impacts while EPA's proposed plan was out for public review. At one event in South Park, John Ryan, a consultant hired by the Lower Duwamish Waterway Group, rose and spoke in Spanish to the largely Latino audience, many of whom were hearing details of the cleanup plan for the first time. He told them that a big cleanup would cause "a line of trucks from here to Olympia," a distance of fifty miles. King County and the Waterway Group also distributed an infographic highlighting the disputed truck and air pollution impacts as fact.[57]

"One thing Ryan did not say explicitly," wrote a reporter who attended the event, "was that he was working for the local governments and the Boeing Co., all of whom as polluters and owners of polluted land are on the hook for [a] share of the cleanup costs." On its face, it appeared that the King County analysis misrepresented the data in order to exaggerate the negative impacts of cleaning up the river and to gain public support for less removal of the river's toxic sediments.[58]

The analysis also suggested that removing contamination would harm fishing communities by stirring up toxic materials in the river's bottom mud, causing the fish to absorb more contaminants than if the chemicals were left undisturbed. But HIA researchers were finding that residual pollution could have serious and permanent impacts on the tribal and subsistence fishing communities that relied on the Duwamish River for their food and cultural needs.

Amber Lenhart, a University of Washington doctoral student, led the health assessment on subsistence fishing communities. She was surprised to learn the importance of fishing to the health of many immigrant and low-income families. "If it's a choice between a free fish or a one-dollar Big Mac, fish is going to have better nutrition," she explained in a 2019 interview. "But the social connections, physical activity, and nature contact are health supporting as well." The study found that trying to prevent people from fishing in the river in order to protect them from illness could eliminate an important food source, increase stress, and worsen their over-all health.[59]

Linn Gould, who studied the cleanup plan's potential impact on local tribes, found that the definition of health in tribal communities differs markedly from the standard medical descriptions typically used by EPA to evaluate risks: preventing disease and having access to clean food and water. Tribal definitions of health include cultural and spiritual metrics that are threatened by environmental pollution and degradation. These include access to local natural resources, maintenance of cultural tradi-tions, and self-determination. The HIA study concluded: "Restrictions and man-made despoliation violate Tribal fishing rights, which will lead to substantial disempowerment, an established determinant of health." The report warned that tribal members would likely continue to eat local fish, even if EPA's cleanup left them with high levels of contamination.[60]

One tribal member consulted during the study articulated the impor-tance of eating salmon. "It's our spiritual food so it feeds our soul," he explained. "It might poison our body, but then we'd rather nourish our soul."[61]

Perhaps the most surprising findings of the HIA, however, were the potential effects of EPA's cleanup plan on the employment prospects of the Duwamish Valley's workers. Employment is strongly correlated with health, and Dr. Bill Daniell found that the financial uncertainty and fear of liability associated with contamination that might be left behind could act as a drag on the local economy and lead to the loss of jobs.[62]

Previous King County reports, which had not been publicly released, concluded that although some jobs might be lost because of cleanup costs borne by local companies, employment opportunities could actually increase. Daniell drew on these reports in his findings: "Existing businesses and employment could benefit substantially," he wrote, "if the cleanup reversed the constraints and stigma of a blighted river and if this stimulated industry revitalization and economic robustness." In other words, although money spent on cleaning up the river would represent a cost to the responsible parties, it might also stimulate local economic growth.[63]

Daniell's assessment also pointed to the benefit of jobs generated by the cleanup itself. One business on the river that benefited from the early action cleanups was the LaFarge Cement Plant, just south of Kellogg Island. LaFarge is the successor to the Holnam Cement Company, the factory sued for its polluting releases by Wingard's Waste Action Project in the 1990s. Built in the 1960s on the historic airfield used by Boeing and US Postal Services, the plant had not changed significantly in nearly fifty years. In 2010, changes in the industry led to the shutdown of LaFarge's enormous kiln, which meant laying off half of the plant's employees unless the company could find a new line of business.[64]

"The plant was dying," said the plant manager, Jonathan Hall, in 2019. If it wanted to "do something with this facility and the people who work here," the company would need to reinvent itself. It now uses its waterfront berths and industrial cranes to serve as an offloading, dewatering, and transfer facility for contaminated sediments dredged from the Duwamish River. Hall describes the change as an "upmarket solution" and says one-third of his work force—about twenty employees—is now involved in cleanup-related business at the plant. The King County internal report cited by Daniell estimated that the full river cleanup would generate 270 year-round jobs and another 210 seasonal jobs during the life of the project, feeding $377 million into the local economy.[65]

The HIA team developed a list of recommendations to the local government agencies responsible for the cleanup (the city, county, and port), as well as to EPA and the state agency responsible for controlling ongoing sources of river pollution. The HIA report was a road map, in effect, for policymakers interested in using the cleanup as an opportunity to improve the overall health of the Duwamish Valley—exactly the same goal first articulated by the community in its Duwamish Valley vision document nearly five years earlier.[66]

A QUESTION OF JUSTICE

Regarding the Duwamish River Superfund cleanup, King County and the University of Washington had dramatically different approaches to questions of health. Both sought to provide information to guide EPA's cleanup plan, but their conclusions were as divergent as their research methods. In the midst of these conflicting studies, EPA decided to conduct its own analysis. As a federal agency, theirs was based on US government policy: in this case, Executive Order 12898, known as the environmental justice or EJ order.[67]

The order, signed by President Bill Clinton in February 1994, stated that the pursuit of environmental justice was part of the mission of every federal agency. It required all agencies to identify and address the environmental and health effects of their policies, programs, and activities on minority and low-income populations. Environmental justice was defined as the "fair treatment" and "meaningful involvement" of all people regardless of race, color, national origin, or income in the development, implementation, and enforcement of environmental laws, regulations, and policies.

The Obama administration had recommitted to the EJ executive order in anticipation of its twentieth anniversary in 2014. Taking a more "targeted and energetic approach" to integrating environmental justice concerns in federal decisions, Obama's EPA provided guidance on how to analyze impacts on minority, low-income, and Native communities and set out a road map for fully implementing the order by 2014. It was in this context that EPA's Duwamish team undertook the first environmental-justice analysis of a Superfund cleanup plan. As with the HIA, its purpose was to determine the likely health impacts of its cleanup proposal and to develop recommendations to minimize or mitigate any adverse or unjust consequences.[68]

The Duwamish Valley's communities had long clamored for attention to what they characterized as environmental injustices. From too much pollution to too few trees, people living in the Duwamish Valley felt over-burdened and underresourced. They also suspected that these conditions were permitted, or even encouraged, to exist because their neighborhoods were poorer and "browner" than the centers of financial and political power in Seattle. Until EPA took on an environmental-justice analysis of its proposed cleanup plan, however it had not formally recognized the people living and fishing in the Duwamish Valley as covered by the order. The analysis established the EJ status of the river's communities.

One focus of the EJ analysis was a set of restrictive covenants and behavioral guidelines, called institutional controls, intended to protect people from exposure to harmful chemicals in their environment. The fence at Terminal 117 that EPA would have required the port to install around the property is one example of an institutional control: it was intended to protect the community by excluding people from the contaminated site. Another is the fishing advisory that had been in place for the Duwamish River since the site was listed. Both kinds of controls put the burden of avoiding the remaining pollution on the affected community rather than on the responsible polluter.[69]

Institutional controls are intended to be in effect only until a site is cleaned up, but in the case of the Duwamish River, EPA was anticipating permanent fishing advisories if ongoing pollution could not be reduced enough to make the river's fish safe to eat. In that case, the EJ analysis said, EPA needed to consider how to mitigate the impact on the river's fishing communities. Instead of simply posting a warning sign, the analysis con-cluded, EPA should require mitigation or compensation for the lost resources—something that had never been done before.[70]

The EJ analysis concluded that fishing advisories often wrongly assume there are accessible and affordable food substitutes for fishers, and that attempting to change people's behavior is an acceptable approach. In addi-tion, the EJ analysis found that "restrictions on fish consumption may also lead to short- and long-term changes in diet with significant health consequences."[71] The Administration for Native Americans, an office under the federal Department of Human and Health Services, agreed, calling fishing restrictions on the Duwamish an "untenable burden for subsistence and tribal fishers."[72]

A Washington Health Department sign warning people to avoid eating fish from the Duwamish River that are contaminated with toxic chemicals. Photo by the author.

The report also warned that discouraging people from eating fish "may be akin to recommending abandonment of their cultural heritage and identity." James Rasmussen, who became director of the DRCC in 2010, often spoke about the significance of a healthy river to his own Duwamish Tribe. "I am a native of this river and this river is a part of me," he explained. "The species that live here—they are my cousins and my aunties and my uncles and my grandfathers and my grandmothers. They are a part of my family."[73]

The analysis suggested mitigation options such as underwriting the cost of transportation to healthier fishing locations, providing uncontaminated seafood or monetary compensation for lost resources, and developing aquaculture facilities where people could raise and catch their own fish. In all cases, the analysis concluded, any mitigation needed to be decided on in close consultation with the affected communities. This was not a prescription to do something *for* the affected communities, but to do it in partnership with them.[74]

Clifford Villa was EPA's assistant regional counsel while the EJ analysis was being conducted. "The idea that some people should be compensated for the loss of fish—or any natural resource—isn't hard to understand," he said several years later. The tricky part was figuring out who was entitled to compensation and how to do it. Regardless, he said, the needs of the Duwamish community and EPA's EJ analysis "succeeded in putting some very challenging questions of environmental and economic justice on the table."[75]

The EJ analysis was released in conjunction with EPA's proposed cleanup plan in February 2013. Both documents were considered drafts that would be approved only after a public review and consideration of all comments received. The findings and recommendations of the analysis, however, helped to transform the community's expectations about what would happen to them once EPA finalized its cleanup plan and issued its record of decision.[76]

HOW CLEAN IS CLEAN?

The EPA project manager Allison Hiltner explained the details of the agency's proposed cleanup plan at a public meeting at South Seattle College's Georgetown campus in April 2013. When the floor was opened for public testimony, Ken Workman, the four-times great-grandson of Chief Se'alth, began to speak. Alternating between English and the Lushootseed language, he welcomed the attendees to his family's traditional land. He spoke about participating in the University of Washington's Health Impact Assessment on behalf of the Duwamish Tribe, and about being taught by his father not to eat certain fish from the river, because of the cancer in them.

Workman is not what many people in Seattle envision when they think of Native people. A passionate motocross racer and a Boeing Company analyst, Workman began to teach himself Lushootseed at the age of fifty-six. He grew up knowing his family's Duwamish heritage, but he discovered that he was a direct descendant of Se'alth only when he was in his fifties. For him, being Duwamish meant he was "imprinted" on the land around the river. "I was born and raised in West Seattle, on the beach of

Alki, and in the woods along the West Bank Duwamish River," Workman later said. "The trees called. I played in the woods and swam in the river, and this is all I know. Rivers draw my family."[77]

At the meeting, Workman had a direct request for EPA. "I would ask that we be able to use the river, to walk on those beaches without fear, to drop a line in the water and pull up *a bottom fish*, a bottom fish, and throw it on the fire and cook it without fear," he said. "I would encourage that, when we clean up the Duwamish River, that we do it right, we do it for the long term."[78]

Workman's statement marked the first time EPA had received public testimony in a Coast Salish language. As the deadline for a cleanup decision approached, Native and immigrant community leaders, fishermen, business owners, longshoremen, environmental and neighborhood organizations, artists, media personalities, and even the hip-hop icon Macklemore lent their voices to a River for All campaign that sought to strengthen the final plan and protect the health of the river's most exposed and vulnerable constituents.[79]

By the time the public comment period on the plan was over, EPA had received more than 2,300 comments on its proposal in ten languages. The Duwamish River cleanup would affect the health and well-being of dozens of racial, ethnic, and linguistic communities. Those most acutely affected would be people whose health, for economic or cultural reasons, was inextricably linked to the health of the river's fish. These communities— traditionally the most marginalized in highly stratified Seattle—had become the driving force behind the Duwamish cleanup.[80]

The central question EPA had to decide before finalizing a plan for the river that would protect both the environment and people's health was "How clean is clean?" Since the 2001 Superfund listing, federal, state, and local governments had been working to investigate not just what pollution was already in the river, but also its past and present sources. PCBs were ubiquitous; even areas with no direct sources collected PCB particles from the air and from decomposing salmon contaminated with the chemicals. This unnatural yet ubiquitous background level of contamination is high enough to threaten the health of anyone who eats large amounts of fish from the river. Yet this level was the best EPA could hope for, given the global sources of pollution that find their way into Puget Sound, tainting even its most remote rivers and bays.[81]

The bigger problem, however, was the volume of PCBs continuing to run into the river from sources within its watershed. If left uncontrolled, continuing pollution would stall cleanup of the river with PCB levels far above safe or legal limits. Where most of the PCBs were coming from, however, remained a mystery, one that threatened to undermine the effort to protect the health of the river and its people.

The Lower Duwamish Waterway extends five and a half miles from the southern tip of Harbor Island, on the river's former tide flats. It collects runoff from twelve square miles of urban and industrial lands, three highways, the King County Airport, and the CSOs that route sewage and stormwater from a much larger area to outfalls on the river. It also receives all of the water from the upper Duwamish and Green Rivers, an area of 480 square miles.[82]

Following construction of the Howard Hansen Dam on the Green River, extensive tracts of farmland were replaced with industries, shopping malls, and parking lots, all of which introduced additional chemical pollution into local streams and the Duwamish River. The consultants hired by the Lower Duwamish Waterway Group stated that no matter how much polluted bottom mud was removed, pollution levels in the river would ultimately depend on how much pollution was still coming in from upriver.[83]

Technical consultants working for the DRCC disagreed with the Waterway Group's assumptions. They countered that the more toxic mud was left behind, the greater the risk that it, too, would add to the river's pollution burden as it was, inevitably, reexposed—scoured by ships, disturbed in an earthquake, or simply churned up during the increasingly violent floods that were accompanying climate change. Removing more pollution would result in a cleaner river and a greater likelihood that the river would stay clean in the long run.

The Waterway Group and the DRCC also disagreed on what could— and couldn't—be done to reduce upriver pollution. The Waterway Group wanted required cleanup levels to be based on its own estimates of the amount of pollution coming from upriver sources before cleanup began. As far as they were concerned, upriver pollution was simply a preexisting condition that constrained what could be achieved by the cleanup. Their estimates were based on a model; real data were scarce. The model did not assume upriver pollution would be reduced, and the group did not believe that EPA had the authority to require upriver pollution controls in their cleanup order. The DRCC, by contrast, advocated for controlling upriver pollution before cleanup. Only once those sources of pollution were found and controlled, the coalition argued, could EPA know how clean the river could and should be.[84]

In 2007, after lengthy debates, the Washington State Department of Ecology stepped in. Their focus up to that time had been on finding and preventing ongoing pollution within the Duwamish Superfund site—itself a herculean task. They now decided to look into just how much pollution might actually be coming from upriver.[85]

To identify pollution sources, the state took sediment samples from the upper Duwamish and Green Rivers. If the concentrations were the same throughout the upriver area, that would indicate ubiquitous sources throughout the watershed—which could confirm the Waterway Group's assumptions. Isolated hot spots of chemical contamination, however, would indicate a localized source that could be identified and controlled. Initial tests suggested that contaminated hot spots of PCBs and other chemicals were tied to particular locations and sources. The state's data also showed that upriver pollution loads, particularly for key chemicals like PCBs, were not nearly as high as the Waterway Group had estimated.[86]

Once the ecology department began investigating, King County also began testing water and riverbed sediments upriver to determine where upper Duwamish and Green River pollution was coming from. While the Waterway Group publicly stated they would not do any work upriver of the Superfund site, Dan Cargill, the state's source control manager for the Superfund site, credits King County with "taking a longer, more comprehensive view of things." From the beginning, the pipes and industrial lands east and west of the lower Duwamish River were recognized as sources of pollution that should be controlled before the cleanup was done. Upriver pollution to the south, however, had been considered outside the scope of the Superfund investigation, even though the lower river was a sink for sources in every direction.[87]

Before EPA issued its final cleanup decision, the river's responsible parties took the position that they were not liable for the costs of stopping upriver pollution. King County in particular, despite its new sampling effort, dug in its heels: since the upper watershed was county land, they likely faced the greatest costs if upriver pollution sources to the Duwamish had to be controlled. But they also had the most at stake if it continued unabated. The county was acutely aware that the entire Duwamish River was an active fishing ground for the Muckleshoot Tribe. Faced with the argument that the tribes had treaty fishing rights that needed to be protected from pollution, one King County staffer protested: "The Tribes have rights to fish, but not to clean fish."[88]

Five years before EPA issued its final cleanup decision, Hedrick Smith of *Frontline* had interviewed the state governor, Christine Gregoire. "How clean is clean?" he asked her. "Do we need to get rivers back to where people can fish and safely eat the fish without fear to their health?" Gregoire had responded: "That's the goal. That *is* the goal. That *has* to be the goal."[89]

In the end, EPA concluded that no amount of PCB contamination in the river sediment was safe or acceptable. In its final cleanup order, issued on November 21, 2014, EPA set a target of 2 ppb for PCBs in sediment—the background level found in the most pristine parts of Puget Sound. In an accompanying plan, the state ecology department committed to overseeing pollution control efforts along the Duwamish Waterway and upriver. Only after a concerted effort to clean up the river and all contributing pollution throughout the watershed would the state and federal agencies consider whether the river had reached an "equilibrium" that justified relaxing EPA's cleanup standard. Thirteen years after the Lower Duwamish Waterway was listed as a Superfund site, the cleanup plan for the whole river was finally done. The work could start just as soon as ongoing pollution sources were controlled enough to protect the river from being recontaminated. Once started, EPA estimated the cleanup would take seventeen years to complete, at a cost of nearly $350 million.[90]

A human figure hovered, suspended, inside a white translucent sphere seventy-five feet above the river, a vision of delicacy and grace despite the crane rigging and arc of cable preventing it from plummeting into the black water below. The figure cast a shadow on the paper globe tethered to the crane's neck as it danced within, appearing half human and half pupa as it slowly descended and emerged from the illuminated orb into the night air. Flowing fabric attached to the dancer's arms and legs completed the illusion of gestation.

Tanya Brno was performing the most unusual aerial dance of her career, one that included literally walking on water. The piece, with music accompaniment by the Native flute player Paul Cheoketen Wagner, was part of a summer-long arts and culture program called Duwamish Revealed. In 2015, a year after the river's environmental fate had been decided by EPA, different forces were beginning to shape the river's social and cultural place in the future of Seattle. Artists were bringing life to a message from the Duwamish Valley community: "We're tired of fighting. We want to celebrate."[91]

As Brno descended headfirst toward the river, the smokestack at LaFarge Cement loomed as a stark industrial backdrop to the display of limbs and silky gauze. LaFarge was a partner in the arts project. Pacific Pile and Marine, an industrial contractor based in South Park, donated both the use of the crane and the labor of its skilled operators, who guided Brno through the air and across the surface of the water.

The Environmental Coalition of South Seattle (ECOSS)—one of the Duwamish River Cleanup Coalition's founding members—served as the artists' sponsor and project manager. The city, the county, the Port of Seattle, and Boeing all provided grants or material support for the program. The fifty-plus installations and live performances that made up the program represented the first expression of cooperation and collaboration between factions that had spent more than a decade arguing about the river cleanup.

The local arts critic Jim Demetre described Duwamish Revealed as a celebration of the river's "thriving and tragic past, complicated present and transformative future. . . . The extent of support from municipal agencies, environmental groups, cultural organizations and local businesses bodes well for future joint efforts at the river. Together they've opened a public arena where a discussion of the river's future can take place." The organizers, Nicole Kistler and Sarah Kavage, hoped to help transform the river from toxic blight to public asset, bringing their own artwork as well as their expertise in urban planning and landscape architecture to the project. "In the making, in the sharing, in the dialogue there's a lot of healing both for the place and for the people experiencing the place," said Kistler.[92]

The project also attracted attention from people who knew little about the river or even about the history of Seattle. "The art creates a reason to come down to a place that most Seattleites haven't seen before," said Bill Pease, a South Park resident who works at ECOSS. At the opening ceremony, James Rasmussen told the audience: "This is home. It's important that you understand that you are in a place that has been a home for people for over ten thousand years. You all have only been here for a little over a hundred-and-some-odd years? God damn, you can change a place."[93]

Kavage and Kistler wanted to encourage not only river-inspired artistic expression but also creative expression by the river's diverse Native and immigrant cultures. "'Cultural affirmation is the highest form of empowerment,'" Kavage said, attributing the statement to the Duwamish Revealed artist Jose Montano. "That has become a mantra for us."[94]

In one of the festival's ceremonial performances, Sophorn Sim knelt by the river's edge and released a small, neatly folded boat of banana leaves into the water. Lit candles sat atop the boat as it drifted into the current, joining several dozen small, illuminated rafts floating away from the park. It was twilight, at the end of the Water Festival in South Park, an event filled with performances by Cambodian, Filipino, Vietnamese, and Mexican artists. The traditional Cambodian lantern ceremony, known as Loi Protip, was being held just downhill from White Center, home to Seattle's largest Cambodian community.

Next to Sophorn stood Song Vann, a Buddhist monk who was guiding the participants. He had not performed the ceremony in over forty years: it had been all but lost with the flight of Cambodian refugees from the labor camps of the Khmer Rouge several decades earlier. In Cambodia, performing Buddhist ceremonies had been punishable by exile or death. An estimated fifty thousand Buddhist monks were executed during the Khmer Rouge regime from 1975 to 1979. Vann was one of the survivors.[95]

All afternoon, children and adults had been making the floats that would carry the candles downriver, along with people's prayers. Ruben Chi, whose mother immigrated to the United States from Venezuela, launched one of the floats and watched as it was carried away with the outgoing tide. He explained that its purpose was to "give gratitude to nature and the Duwamish River, Seattle's only river."[96]

The revitalized lantern ceremony was just the most recent of Sim's labors of love for Seattle's immigrants and refugees. "When Cambodia was under the Rouge regime for three years, eight months, and twenty days—I remember exactly—all religious and cultural festivals were banned," she said in 2019. Sim herself was only nine years old when she was sent to a child labor camp by the Khmer Rouge. When Vietnam invaded Cambodia in 1979, she and her family escaped through minefields to a refugee camp on the Thai border. She was eventually admitted to the United States as a refugee in 1985, arriving in Seattle in the fall of that year.[97]

For those like her who left Cambodia, "the lantern festival was all but forgotten," she said, even though "the river back home was very precious to our people." They didn't know what the rules were in their new home, and some were still afraid. When ECOSS and the arts festival gave her an opportunity to re-create the festival of Loi Protip on the Duwamish River, Sim was overjoyed. "The lantern ceremony has been awakened by that," she said, noting that the Cambodian community has asked that the ceremony be continued in future years. "The culture that is embedded in our people for a long time is to celebrate and give thanks to Mother Nature. We want to teach the young generation to appreciate the environment surrounding us, that we depend on."[98]

To Sim, culture is also a bridge to improving the environment and well-being of her Cambodian neighbors. Working as a community health advocate, Sim educates Cambodian fishermen about the risks of eating fish and shellfish from the Duwamish River. She suggests healthier options, like fishing at sites away from the river or consuming only less-contaminated fish, like salmon. But fishing "is their culture," Sim says.

"They are not fishing as much as they do in their home country or town, and when they have the river right next door, they try to fish. They are not necessarily aware of pollution."[99]

In March 2019, Sim participated in a storytelling event organized by Sarah Kavage at which she and other Native and immigrant community leaders in the Duwamish Valley expressed the centrality of rivers to their own personal and cultural experience. Sim sang a Cambodian song she wrote for the occasion. "Rivers in my hometown were so precious and important to us," Sim said. "No matter what part of the world you have come from, somewhere, somehow, I believe you connect with your river."[100]

Duwamish Revealed was not the first creative celebration of the river. Belulah Maple, born in 1893 to one of Jacob Maple's daughters, drew scenes of the Duwamish River and the settlers' early imprints on its landscape. Today prints of her work are displayed at the Tukwila Heritage and Cultural Center, managed by Louise Ann Jones, a Maple descendant herself. Belulah did not consider herself a professional artist; she began producing illustrations in her later years, when she saw that her grandchildren were forgetting the stories of her family's early life on the river. "I'm trying to put our history back where it belongs," she told the *Seattle Times* in 1967. "My artwork may be amateurish, but I think it gets across what I'm trying to say."[101]

Born thirty years after Belulah Maple, Richard Hugo of White Center wrote poems about his dysfunctional family life and the river where he found solace. His first book was published in 1961, while he was working as a technical writer for Boeing. It included the poem "West Marginal Way," which describes the river he knew as a young man. "One tug pounds to haul an afternoon of logs upriver. The shade of Pigeon Hill across the bulges in the concrete crawls on reeds in a short field, cools a pier and the violence of young men after cod."[102]

In 1998, the Seattle artist Gene Gentry McMahon began to explore the Duwamish River and images from its history, including a piece inspired by Jacob Maple's twin granddaughters. Her earliest piece, "Cry Me a River," was exhibited at the Seattle Convention Center in 2008. An eight-by-six-foot mural illustrating historical views of Puget Sound and the Duwamish estuary was displayed at the Seattle Aquarium in 2011. Other artists followed suit, including the Mexican-born Amaranta Sandys, the Swiss transplant Fiona McGuigan, and the Chinese American painter Juliet Shen. A trio of experimental artists known as Sutton-Beres-Culler lent their name and imagery to the River for All campaign in 2014, as did Macklemore and the Native glass artist Preston Singletary. South Park and Georgetown have both sprouted vibrant community-based arts organizations, and a Duwamish River artist residency, cofounded by McGuigan and fellow artist Sue Danielson, has been meeting along the river for ten days every summer since 2012.[103]

Yet the history of immigrant artists pales in comparison to that of the river's Native artisans, dancers, and storytellers, for whom the Duwamish River has long been a subject of artistic and cultural expression. In addition to Singletary's work, a striking contemporary example is a cedar tile collage representing a traditional ceremonial basket, which serves as a centerpiece on the floor of the Duwamish Tribe's cultural center, located along the river bend that Cecile Hansen, the Duwamish tribal chairwoman, saved from development. Designed by the Native artist Mary Lou Slaughter, a basket weaver and a direct descendant of Se'alth, it was installed for the center's grand opening in 2009. James Rasmussen of the DRCC, a jazz artist, performed at the center in 2015 with the composer and musician Steve Griggs, playing a series of compositions called *Listen to Seattle*, based on the life and words of Chief Se'alth.[104]

The many works of art that have been inspired by the Duwamish River's history and its changing character remind us that the restoration and cleanup of the river is not just a scientific or policy question but one of heart and culture as well. Today, seven generations after Rasmussen's and Slaughter's ancestors welcomed the first immigrants to the watershed, a new generation of artists and advocates are taking their place in the river's history.

GENERATION NEXT

The DRCC's Paulina Lopez gathered a group of three dozen middle- and high-school students at Duwamish Waterway Park on the afternoon of March 30, 2019. They assembled at the edge of the park, getting ready to welcome friends, family, and elected officials to the unveiling of their latest project, a celebration of the river's past and their vision for the future. One of the teens slung a guitar over his shoulder, and the group began to walk, singing, across the manicured field toward the former electric company warehouse at the park's south end.

Lopez joined the DRCC staff in 2012 to help with the creation of the Duwamish Valley Vision. She arrived in the United States from Ecuador at age twenty-six and lived in Miami while completing a degree in international human rights law. She moved to Seattle with her husband in 2004, when she was pregnant with their first child. She was drawn to South Park by its vibrant Latino community and its proximity to the Duwamish River. When Lopez first arrived, she did not know that the beach park a few blocks from her new home was not a safe place for children to swim or fish.[105]

A year after she arrived, a neighbor's teenage son was fatally shot. That shook Lopez, who became involved in seeking ways to help solve the problems threatening the community's youth, including family poverty, incarceration, gang violence, and a lack of structured youth programming and gathering places. Ten years later, with the DRCC's support, she started a new program, the Duwamish Valley Youth Corps, to serve the neighborhood's middle- and high-school age children, focusing on ways to make their community cleaner, healthier, and safer. The program thrived. "And we haven't stopped," Lopez said in 2019. "In five years, we've engaged six hundred youth."[106]

At the start of its second year, the Youth Corps hired Carmen Martinez, a veteran youth coach, to help manage the program and its growing number of teens. Martinez is a third-generation South Park resident whose grandmother moved to Seattle in the 1950s from near the Texas border with Mexico. Her family was part of the first wave of Latino migrants to the Duwamish Valley. As a child in the 1960s, Martinez watched her uncles swimming in the river. "They would get sores," she recalls, "but they'd still do it." Together, Lopez and Martinez set out to empower the neighborhood's youth, teaching them about the river, environmental justice, and the issues affecting their community's well-being.[107]

Paulina Lopez, a South Park resident originally from Ecuador, visiting a Duwamish River street-end park with her children in 2013. Courtesy of Paul Joseph Brown.

As Lopez and the teens gathered at the river, they talked about what it meant to welcome their guests to this place. Usually, on a sunny Saturday like this one, the group could be found building rain gardens to help clean the neighborhood's stormwater of its pollutants or planting new habitat for fish and wildlife along Hamm Creek or the Duwamish riverbank. Many of the teens who participated in the Youth Corps went on to take paid positions with the Duwamish Infrastructure Restoration Training (DIRT) Corps—a green-jobs incubator and professional development program for young adults. Environmental restoration is a core component of the Youth Corps curriculum. But today, the group was preparing to unveil a mural they had painted on the wall of the old warehouse next to Duwamish Waterway Park.

Over several weeks, the Native artist Roger Fernandes had helped the group design the Salish-inspired mural. The first section of the mural depicted the natural river bends of the past and Native stories about the origin of the river, as well as a ten-foot-tall image of a Duwamish man

Members of the Duwamish Valley Youth Corps gather to promote the River for All campaign at Duwamish Waterway Park in South Park in 2013. Courtesy of the Duwamish River Cleanup Coalition.

with his arms raised in a gesture of welcome. "It was important to the youth that we portray him as welcoming his visitors," Fernandes said. Most of them, after all, had come from distant places themselves, including Mexico, Honduras, and Somalia. The Duwamish tribal elder Ken Workman had spoken with the group as they embarked on the project and made a point of raising his arms to them to let them know that they were welcome here. They wanted to share that welcoming gesture with other immigrant and refugee families living nearby.[108]

The second section of the mural, still unfinished, would depict the current Duwamish River flowing under a wide bridge. The historic Fourteenth Avenue Bridge connecting South Park to the rest of Seattle had been closed between 2010 and 2014 while city and county leaders debated whether to rebuild it. Neighborhood advocates eventually prevailed in getting the bridge replaced, but the closure and prolonged uncertainty had hampered transportation and emergency services and left South Park feeling once again like the city's neglected stepchild, in the words of Tony Ferrucci a generation earlier. "Imagine what it meant to them," mused Workman, standing at the back of the crowd. "You are marginalized. You don't matter." The bridge in the mural represents the community's triumph in demanding to be heard and to be treated as a priority by their local governments.[109]

The third section of the mural showed tall trees and fish along the river. Fernandes explained, "They want it to be clean again. They can swim in the river. They can canoe in the river. They can fish from the river. All the things they should be able to do, they want to see in their future."[110]

After the teens unveiled the mural, the Duwamish Tribe's chairwoman, Cecile Hansen, congratulated them on their efforts, and Lopez introduced the Port of Seattle commissioner Ryan Calkins. The port had funded the mural and was planning a paid internship program for Duwamish Valley youth. "At the outset, the Port of Seattle was about bending this river to our will," Calkins said. "Today, we understand that was a mistake. Now we understand that the way we ought to care for it is by thinking of the river seven generations ahead." If Calkins's thinking prevails in guiding the port's activities, the effect could be transformative for the Duwamish River.[111]

Concluding the unveiling ceremony, Roger Fernandes once again addressed the crowd, asking for their help in completing the final phase of the mural. "We want to make sure that on the future side, we add some cultural imagery that comes from you, of the cultures that brought you here." Speaking for the Youth Corps, Daniella Cortez agreed. "Creating that bridge between ages, between races, is going to be amazing for our city."[112]

ALL TOGETHER NOW

By 2018, cleanup of the early-action areas that EPA had identified in 2004 had removed about 50 percent of the PCBs from the river bottom. Design and engineering for the riverwide cleanup were also well under way. In early 2019, the Puget Soundkeeper Alliance reached a \$1 million settlement agreement with an industrial metals recycler in the center of the Superfund site to eliminate continuing PCB pollution. Other sources of pollution from the surrounding Duwamish Valley and upriver were gradually being identified and brought under control. Perhaps most significantly, EPA's environmental justice analysis had spurred the agency to consult directly with subsistence fishers on how to conduct the cleanup. They were working on strategies to protect fishing families without putting unjust burdens on those who were not responsible for the pollution.[113]

Beyond the scope of the Superfund cleanup, the City of Seattle had established a Duwamish River Opportunity Fund to help carry out the improvements recommended by the University of Washington's health impact assessment. In 2018, the fund had disbursed more than $1 million in grants to community projects. The city also hired the DRCC's former program manager, Alberto Rodriguez, to develop a Duwamish Valley action plan to deliver measurable improvements in the community's health.[114]

The Port of Seattle and King County were also making their own investments in improving environmental and human health in the Duwamish Valley. The worst of the conflicts between local governments and the valley's communities seemed to be fading, with artists and the valley's youth helping to bridge past divides.

James Rasmussen, director of the DRCC, didn't think these initiatives were enough. He envisioned a partnership made up of all of the river's engaged parties—agencies, industries, tribes, environmental organizations, and community members—to promote cooperation and problem solving as the cleanup progressed. The job of the group would not be to determine what should be done to clean up the river—that had already been decided—but to decide *how* it should be done to mitigate adverse impacts and support each group's needs. Rasmussen firmly believed that cleaning up the river was good not just for its fishing and residential communities, but also for its industries—if it was done right.[115]

After the EPA cleanup decision was finalized, Rasmussen spoke at a Green-Duwamish Watershed Symposium attended by community, business, and government representatives in 2016. "Coming out of this, we go back to work or we go back to war," he told them. "Everybody in the room wants the best. Partnering is how we do that."[116]

James Rasmussen welcomes a tribal canoe carrying Chief Seattle's great-great-great-great grandson Ken Workman to the Duwamish Waterway Park during the 2018 Duwamish River Festival. Courtesy of Tom Reese.

EPA was thinking along the same lines. The 2013 environmental justice analysis recommended that EPA develop strategies for enhancing communication and coordination among all parties during the cleanup. In 2017, EPA's community involvement coordinator, Julie Congdon, proposed convening a roundtable discussion to help guide the cleanup work and prevent the kind of disputes over conflicting versions of technical and scientific facts that had plagued earlier cleanup planning. The participants would seek ways to mitigate impacts and benefit affected communities as the cleanup was under way.[117]

If the roundtable approach works, it will bring together everyone with a stake in the river cleanup for the first time. Although the fundamental requirements of the cleanup have been determined by EPA, details like timing, technology, and mitigation measures could still make or break the Duwamish River Superfund project.

Jonathan Hall, the LaFarge Company's plant manager, was among the industrial representatives at the preliminary convening of the roundtable in 2018. LaFarge is liable for costs related to its own pollution of the river and is also one of the companies benefiting from cleanup-related income and job creation. He acknowledged that the company had vested interests in the cleanup, "but ironically these interests can be in conflict with each other." He suggested that the dual roles of responsible party and cleanup contractor gave his business a valuable "yin and yang" perspective on the cleanup and the complexity of issues it hopes to address. "In short, my involvement is a genuine interest in helping to make the process go more smoothly for everyone," he said after the initial gathering.[118]

Alberto Rodriguez described the roundtable as an opportunity to "co-create" solutions and build trust among all of the river's disparate interest groups.[119] James Rasmussen leads the roundtable's community advisory group caucus. "We have only asked for an open and transparent process with everyone at the table. Now we have it, and it's up to us to make this work for everyone," Rasmussen said. "We have learned that if given a small chance to lift the community voice, it will not only be heard, but will move mountains."[120]

More effort will be required to ensure inclusion of the river's most heavily affected communities: no representatives of the tribes or the sub-sistence fishing community were present at the first meeting, but several fishing community representatives have since joined. When and if the roundtable approach is established, Rasmussen hopes it will enable the Duwamish communities—ancient and new, resident and industrial—to work together to shape the future of the Duwamish River for the next generation.

The Duwamish Valley's earliest industry is still thriving in the twenty-first-century. In 2018, at least fifteen breweries were operating in South Park, Georgetown, and the filled tide flats that make up the city's Sodo District. In April 2019, glasses of Counterbalance Brewing's Pale Ale were lined up alongside samples of Fran's Chocolate, a relative newcomer to the valley, housed in the historic Rainier Brewing building in Georgetown.

The Duwamish Valley businesses were supporting a fund-raising event for the DRCC, held at the Duwamish Tribe's Longhouse and Cultural Center. The event also marked a leadership transition: the outgoing director of the DRCC, James Rasmussen, was being succeeded by Paulina Lopez. The significance of a first-generation Latina South Park resident taking over from a Duwamish elder was not lost on the crowd.

Jolene Hass, the daughter of Cecile Hansen, welcomed the guests to the longhouse. Over the shoulders of both Rasmussen and Lopez she draped a blue and black blanket decorated with traditional Coast Salish designs. "This is not a gift," said Hass. "It is meant to show our appreciation. We ask that you use these blankets, that they are not stored away. And that when you use these blankets you feel the appreciation, the adoration, the love that we have for you." She thanked Rasmussen for his years of service—to the river, the tribe, and the community at large. Embracing Lopez, she said, "We welcome her in." The guests erupted in applause.

Rasmussen is continuing his work with the organization with a focus on the roundtable, where he hopes to help transform generations of conflict into consensus about the future of the river. Lopez—with the youth she mentors at her side—is setting her sights even higher. She has a vision for the environmental, social, and economic vitality of the entire watershed—one that includes not just a fishable and swimmable river, but clean air, vibrant local businesses, more parks, and affordable housing.

Together, Rasmussen and Lopez hope they have built the momentum necessary to restore the health of Seattle's hometown river and of its Native and immigrant communities for the next generation of children who will live, work, pray, and play along its banks. Lopez is determined to make sure that that the Duwamish Valley communities have what they need not just to clean up their environment but to remain "healthy, strong, resilient, and in place" while doing it.

Expressing cautious optimism about the future of the river valley, Lopez mused, "The wealth of Seattle was built right on the back of the Duwamish River." She admitted, "Sometimes I feel like we're being crushed. But then I wake up in the morning and I see our youth—how much potential they have." She recounted what one young woman said to an elderly neighbor as they worked together planting trees: "Every time I plant a tree, I feel like I'm leaving my heart. . . . I have already planted forty trees, so that means my heart is in forty different places."

The words struck Lopez as joyful. "Hope is always there," she said, smiling at the sea of faces from the Duwamish Valley's neighborhoods, local businesses, and government agencies. "I feel like we're building a movement."

Afterword

IN JUST SEVEN GENERATIONS—FROM SE'ALTH'S PARENTS TO THE Duwamish elder Ken Workman, and from James Rasmussen's great-great-great grandmother, Tupt-Aleut, to his niece, Christine Nelson—the changes brought by Euro-American explorers and colonists in the Northwest have transformed the Duwamish River and its communities nearly beyond recognition. Yet some of the river's Native people and their kin in the natural world hang on. The salmon, the cedar, and the great blue heron can still be found in and near the river if you know where to look. And Duwamish tribal members today frequently echo their chairwoman's mantra when they remind us, "We are still here."

In 2019, the Duwamish Tribe celebrated the tenth anniversary of their longhouse and cultural center, built on the waterway's sole surviving river bend. Erected more than a century after their last ancestral longhouse was burned down, the center serves as a reminder that Native places and people survive in Seattle. The Muckleshoot Tribe, which absorbed many of the Duwamish people, also remind us of this each fall when they lay their nets out on the river, catching salmon for the tribe and for trading, as they have always done, in the commercial market. Despite all the changes, the Duwamish River, its people, and its salmon are inextricably linked.

As this book goes to print, government, industry, and community representatives working to clean up the Duwamish River are struggling to find common ground. Diminishing federal government support for environmental health protections and an eroding commitment to environmental justice in the Environmental Protection Agency under President Donald Trump are also leading once again to uncertainty about the future of the river cleanup.

In 2019, EPA proposed weakening water quality and cleanup standards for a raft of pollutants. Some advocates worry that this change could undermine hard-won health protections for the Duwamish River. Roundtable participants attempting to overcome past divisions are trying to build trust as they move forward together to address the river's challenges. None of this work is easy, and its success is not guaranteed. But most consider the rewards of creating a new model of collaboration to be well worth the trouble. Ridding the city of the stigma of having one of the nation's most contaminated rivers is a powerful incentive to succeed. For this to happen, everyone will need to be at the table—listening, problem-solving, and lifting their share of the historical and contemporary burden—in order to provide for the needs of the city's diverse Native and immigrant communities in the complex urban and industrial waterscape of Seattle's only river.

NOTES

INTRODUCTION

1 David Munsell, interview with author, June 6, 2017.
2 *Continued Archeological Testing at the Duwamish No. 1 Site (Summary of Report),* Office of Public Archaeology, University of Washington, March 1977; Eric Lacitis, "Artifacts Lost, Says Archeologist," *Seattle Times,* July 14, 1976.
3 Lacitis, "Artifacts Lost."
4 *Continued Archeological Testing.*
5 *Continued Archeological Testing.*
6 Munsell interview.
7 Munsell interview.
8 Sarah Campbell, interview with author, June 29, 2017. Efforts to develop the site continued until at least 1986, according to Campbell.
9 Coll Thrush, *Native Seattle: Histories from the Crossing Over Place* (Seattle: University of Washington Press, 2007), 194–95, 205.
10 David M. Buerge, *Chief Seattle and the Town that Took His Name: The Change of Worlds for the Native People and Settlers on Puget Sound* (Seattle: Sasquatch Books, 2017), 99.

CHAPTER 1: IN THE BEGINNING

1 National Park Service interpretive displays, Mount Rainier National Park, WA.
2 Mike Sato, *The Price of Taming a River: The Decline of Puget Sound's Green/Duwamish Waterway* (Seattle, WA: Mountaineers, 1997), 20–24.
3 Sato, *The Price of Taming a River,* 20–24.
4 Cynthia Updegrave, "Cedar-Sammamish Watershed," HistoryLink, December 30, 2016, www.historylink.org/File/20273.

5 "North Wind and Storm Wind," Coast Salish Villages of Puget Sound: Storytelling Sit es, http://coastsalishmap.org/north_wind_and_storm_wind.htm, accessed March 14, 2019. This story was first published by Arthur Ballard in "Mythology of Puget Sound," *University of Washington Publications in Anthropology,* vol. 2, no. 2 (December 1929); this version was adapted by Tom Dailey.

6 "North Wind and Storm Wind."

7 Brian F. Atwater, Satoko Musumi-Rokkaku, Kenji Satake, Yoshinobu Tsuji, Kazue Ueda, and David K. Yamaguchi, *The Orphan Tsunami of 1700: Japanese Clues to a Parent Earthquake in North America* (Seattle: University of Washington Press, 2011).

8 Cynthia Updegrave, "Duwamish-Green Watershed."

9 David M. Buerge, *Chief Seattle and the Town That Took His Name: The Change of Worlds for the Native People and Settlers on Puget Sound* (Seattle: Sasquatch Books, 2017), 20–21; Ken Workman, interview with author, September 26, 2018; Workman family records, genealogical records provided by Se'alth descendant, in author's possession. Some historians have placed Se'alth's birth elsewhere, including at his mother's village of Stuk; the account here is supported by his descendant Ken Workman.

10 Rasmussen/Nelson family records, genealogical, legal, and property records provided by descendants of Kanum family, 1800s–present, in author's possession; United States, Secretary of the Interior, *Summary under the Criteria and Evidence for Proposed Finding against Acknowledgement of the Duwamish Tribal Organization: Anthropological Technical Report* (Seattle: Bureau of Indian Affairs, 1996), 42; George A. Kellogg, *A History of Whidbey's Island* (Coupeville, WA: Island County Historical Society, 1934), 47; Updegrave, "Duwamish-Green Watershed."

11 Kellogg, *A History of Whidbey's Island.*

12 "George Vancouver, A Voyage of Discovery to the North Pacific Ocean," Center for the Study of the Pacific Northwest, https://www.washington.edu/uwired/outreach /cspn/Website/Classroom%20Materials/Reading%20the%20Region/Discovering %20the%20Region/Commentary/4.html, accessed March 18, 2019.

13 "George Vancouver, A Voyage of Discovery to the North Pacific Ocean."

14 George Vancouver, *A Voyage of Discovery to the North Pacific Ocean, and around the World in the Years 1790–95,* vol. 1 (London: G. G. and J. Robinson, 1798), 254–55. Thirty percent of the population had succumbed to disease before Vancouver's visit. See Robert T. Boyd, *The Coming of the Spirit of Pestilence: Introduced Infectious Diseases and Population Decline among Northwest Coast Indians, 1774–1874* (Seattle: University of Washington Press, 1999).

15 Vancouver, *A Voyage of Discovery,* 259–60; Buerge, *Chief Seattle,* 23.

16 Vancouver, *A Voyage of Discovery,* 261.

17 Vancouver, *A Voyage of Discovery,* 262.

18 Vancouver, *A Voyage of Discovery,* 262–63.

19 Vancouver, *A Voyage of Discovery,* 264.

20 Boyd, *The Coming of the Spirit of Pestilence.*

21 Buerge, *Chief Seattle,* 25.

22 "Fur Trade in Oregon Country," Oregon Encyclopedia, accessed March 18, 2019, https://oregonencyclopedia.org/articles/fur_trade_in_oregon_country; George Dickey, ed., *The Journal of Occurrences at Fort Nisqually: May 30, 1833 to April 25, 1835*, vol. 1 (Tacoma, WA: Fort Nisqually History Site, Tacoma Parks Department, 1989).

23 Buerge, *Chief Seattle*, 37–40.

24 Buerge, *Chief Seattle*, 32–33; "Good Chief Seattle: How a Young Warrior Became Ruler of Many Tribes," *Seattle Post-Intelligencer*, March 26, 1893; Ken Workman interview.

25 "Good Chief Seattle."

26 "Good Chief Seattle."

27 "Good Chief Seattle."

28 Clarence Bagley, *Early Days in Seattle*, vol. 1 (Seattle, WA: Argus Press, 1920), 112.

29 Kenneth D. Tollefson, "Political Organization of the Duwamish," *Ethnology* 28, no. 2 (April 1989), doi:10.2307/3773671; Secretary of the Interior, *Summary under the Criteria and Evidence*, 2.

30 T. T. Waterman, Vi Hilbert, Jay Miller, and Zalmai Zahir, *Puget Sound Geography* (Federal Way, WA: Lushootseed Press, 2001); Coll Thrush, *Native Seattle: Histories from the Crossing-Over Place* (Seattle: University of Washington Press, 2007), 89.

31 Tollefson, "Political Organization of the Duwamish."

32 Rasmussen/Nelson family records; Secretary of the Interior, *Summary under the Criteria and Evidence*.

33 Ann Rasmussen, "Oral History," interview with Arlene Wade, October 4, 1998, Duwamish Tribe Longhouse and Cultural Center; James Rasmussen, interviews with author, 2015–19; Tina Hilding, "Tribe's Past to be Paved," *Valley Daily News* (Renton, WA), August 3, 1990; James C. Chatters, *Archaeology of the Sbabadid Site 45KI51, King County Washington* (Seattle: Office of Public Archaeology, University of Washington, 1981); Hill Williams, "Dig Unearths the Good Duwamish Life," *Seattle Times*, October 7, 1979; Tina Hilding, "Renton Asks State to Protect Artifacts at Building Site," *Valley Daily News* (Renton, WA), August 14, 1990; Coll Thrush, personal communication, March 30, 2019.

34 An earlier settler, Thomas Glasgow, was chased off in 1848 by Natives suspicious of the new white settlers, despite his being married to the daughter of the local leader Patkanim.

35 Rasmussen/Nelson family records; Coupeville History Museum records, Coupeville, WA; Art Gorlick, "A Monument to Indian History: Chiefs Buried at Remote Site on Whidbey Island," *Seattle Post-Intelligencer*, April 26, 1988.

36 Rasmussen, "Oral History."

37 Don M. Carr to the Commissioner of Indian Affairs, February 8, 1916, in Rasmussen/Nelson family records; Rasmussen, "Oral History." Carr was superintendent of the Yakima Agency of the United States Indian Service.

38 Rasmussen, "Oral History."

39 Walt Crowley, "Great Britain and the United States Sign the Treaty of Joint Occupation of Oregon on October 20, 1818," HistoryLink, January 23, 2003, www.history
link.org/File/5103; Dickey, *Journal of Occurrences at Fort Nisqually*.

40 Buerge, *Chief Seattle*, 53–60. Ouvre was the first white man known to have traveled
up the Duwamish River, originally called "Ouvre's River" by the Hudson Bay Company. See Chalk Courchane, "Jean Baptiste Ouvre and His Sons-in-Law, Francois
Xavier Seguin dit Laderoute, Louis Hercule Lebrun, Adolphe Lozeau (L'Oiseau)
and Antoine Gregoire: To the Pacific Northwest in 1810," Oregon Pioneers, www
.oregonpioneers.com/bios/JeanBaptisteOuvre.pdf, accessed October 7, 2019.

41 Buerge, *Chief Seattle*, 56; "Leschi, The Nisqually Chief," Chief Leschi Schools, www
.leschischools.org/Page/103, accessed March 19, 2019.

42 Jim Kershner, "Britain and the United States Agree on the 49th Parallel as the Main
Pacific Northwest Boundary in the Treaty of Oregon on June 15, 1846," History-
Link, July 31, 2013, www.historylink.org/File/5247; J. A. Eckrom, *Remembered
Drums: A History of the Puget Sound Indian War* (Walla Walla, WA: Pioneer Press
Books, 1989), 116; An Act to Create the Office of Surveyor-General of the Public
Lands in Oregon, and to Provide the Survey, and to Make Donations to Settlers
of the Said Public Lands, *United States Statutes at Large* (1850), chapter 76, 496,
www.loc.gov/law/help/statutes-at-large/31st-congress/c31.pdf ; Margaret Riddle,
"Donation Land Claim Act, Spur to American Settlement of Oregon Territory,
Takes Effect on September 27, 1850," HistoryLink, August 9, 2010, www.history
link.org/File/9501.

43 Buerge, *Chief Seattle*, 99; Dickey, *Journal of Occurrences at Fort Nisqually*, sections
6.24, 7.4.

CHAPTER 2: AND THEN THERE WAS BLOOD

1 "Story of a Pioneer Party in King County Forty Years Ago," *Seattle Post-Intelligencer*,
January 21, 1906; C. T. Conover, "Just Cogitating: Pioneer Dinner Party in
Duwamish Valley Described," *Seattle Times*, October 5, 1958.

2 Conover, "Just Cogitating"; Coll Thrush, *Native Seattle: Histories from the Crossing-
Over Place* (Seattle: University of Washington Press, 2007), 20.

3 Edward S. Meany, "Living Pioneers of Washington," *Seattle Post-Intelligencer*,
November 10, 1915; Conover, "Just Cogitating"; Kay Frances Reinartz, *Tukwila:
Community at the Crossroads* (Tukwila, WA: City of Tukwila, 1991), 5–10.

4 Meany, "Living Pioneers"; Conover, "Just Cogitating"; Eli Bishop Mapel, The
Seattle-Maple History Project, MS P-B 66, Bancroft Library, University of California, Berkeley.

5 Clarence B. Bagley, *History of Seattle: From the Earliest Settlement to the Present Time*
(Chicago: S. J. Clarke, 1916), 29, 225; Conover, "Just Cogitating."

6 Conover, "Just Cogitating"; David M. Buerge, *Chief Seattle and the Town That Took
His Name: The Change of Worlds for the Native People and Settlers on Puget Sound*
(Seattle: Sasquatch Books, 2017), 107–9.

7 Conover, "Just Cogitating."

8 Buerge, *Chief Seattle,* 212–14.

9 Buerge, *Chief Seattle,* 94–101; Thrush, *Native Seattle,* 28–30.

10 Edmond S. Meany, "Chief Patkanim," *Washington Historical Quarterly,* 15, no. 3 (July 1924), https://journals.lib.washington.edu/index.php/WHQ/article/view/6581 /5653; J. A. Eckrom, *Remembered Drums: A History of the Puget Sound Indian War* (Walla Walla, WA: Pioneer Press Books, 1989), 109.

11 Eckrom, *Remembered Drums;* Meany, "Chief Patkanim."

12 Thrush, *Native Seattle,* 231–35; "B. F. Shaw Tells an Exciting Story of Indian Days on the Sand Spit," [1904,] Clarence Bagley papers, Box 17, Folder 8, ACC. 36, University of Washington Libraries, American Indians of the Pacific Northwest, https://content.lib.washington.edu/aipnw/my_first_reception_speech.html accessed March 19, 2019.

13 "B. F. Shaw Tells an Exciting Story."

14 "B. F. Shaw Tells an Exciting Story."

15 George A. Kellogg, *A History of Whidbey's Island* (Coupeville, WA: Island County Historical Society, 1934), 17–24; Buerge, *Chief Seattle,* 107–8. Kruss Kanum's Duwamish wife, Tupt-Aleut, may also have been related to Se'alth. A Bureau of Indian Affairs report on the surviving Duwamish families indicates that they shared common ancestry through Se'alth's mother. United States, Secretary of the Interior, Bureau of Indian Affairs, *Summary under the Criteria and Evidence for Proposed Finding against Acknowledgement of the Duwamish Tribal Organization: Genealogical Technical Report,* June 18, 1996, 3.

16 Buerge, *Chief Seattle,* 95–96, 99–101; Thomas W. Prosch, "A Chronological History of Seattle from 1850 to 1897, Prepared in 1900 and 1901," unpublished ms., University of Washington Libraries, Northwest Collection, Seattle, WA, 25–29. Fay later became an Indian agent and settled on Whidbey Island.

17 Arthur Armstrong Denny, *Pioneer Days on Puget Sound* (Fairfield, WA: Ye Galleon Press, [1888] 1979), 35–36; Thrush, *Native Seattle,* 28–29; Bagley, *History of Seattle,* 17–18.

18 A. Denny, *Pioneer Days,* 15–17; Bagley, *History of Seattle,* 26–27.

19 Cynthia Updegrave, "Duwamish-Green Watershed," HistoryLink, December 31, 2016, www.historylink.org/File/20272.

20 Edward Huggins, *Reminiscences of Puget Sound,* ed. Gary Fuller Reese (Tacoma, WA: Tacoma Public Library, 1984), 43–52.

21 Huggins, *Reminiscences of Puget Sound,* 43–52.

22 A. Denny, *Pioneer Days,* 51.

23 United States, National Oceanic and Atmospheric Administration, Office of Response and Restoration, *Pre-assessment Screen: Lower Duwamish River* (Seattle, WA: Elliott Bay Trustee Council, 2009); Clarence Bagley, *History of King County, Washington,* vol. 1 (Chicago: S. J. Clarke, 1929), 100.

24 Bagley, *History of King County,* 98; Mapel, The Seattle-Mapel History Project; Emily Inez Denny, *Blazing the Way: Or, True Stories, Songs and Sketches of Puget Sound and Other Pioneers (1899)* (Whitefish, MT: Kessinger, [1909] 2011), 63.

25 Mapel, The Seattle-Mapel History Project.

26 Prosch, "A Chronological History of Seattle," 41, 55; David M. Buerge, *Renton: Where the Water Took Wing* (Northridge, CA: Windsor, 1989), 10.

27 Buerge, *Renton*, 10; James Rasmussen, interviews with author, 2015–19.

28 Priscilla McLemore, Descendants of Mary Kennum, Genealogical record of descendants of Mary Kennum aka Tyee Mary or Tupt-Icut, Duwamish Tribal Office, Seattle, WA; Rasmussen/Nelson family records, genealogical, legal, and property records provided by descendants of Kanum family, 1800s–present, in author's possession.

29 Prosch, "A Chronological History of Seattle," 55; Laura McCarty, "Coal in the Puget Sound Region," HistoryLink, January 31, 2003, www.historylink.org/File/5158.

30 Bagley, *History of Seattle*, 127.

31 Bagley, *History of Seattle*, 100.

32 "Poor Old Angeline: Only Living Child of the Great War Chief Seattle," *Seattle Post-Intelligencer*, August 2, 1891.

33 Thrush, *Native Seattle*, 231–40.

34 Cynthia Updegrave, "Cedar-Sammamish Watershed," HistoryLink, December 30, 2016, www.historylink.org/File/20273; Isaac N. Ebey to Michael T. Simmons, September 1, 1850, describing the geography of the Puget Sound region, September 1, 1850, Pacific Northwest Historical Documents Collection, University of Washington Digital Collections, https://digitalcollections.lib.washington.edu/digital/collection/pioneerlife/id/3999/rec/1.

35 David Wilma, "Luther Collins and Two Others Lynch Masachie Jim near Seattle on July 15, 1853," HistoryLink, August 29, 2001, www.historylink.org/File/3525; A. Denny, *Pioneer Days*, 60.

36 Eckrom, *Remembered Drums*, 5; Buerge, *Chief Seattle*, 133–46; Thrush, *Native Seattle*, 50–51.

37 George Gibbs, *Report on the Indian Tribes of the Territory of Washington* (1854), cited in Bureau of Indian Affairs, *Summary under the Criteria and Evidence*, 18; T. T. Waterman, 1920, cited in Bureau of Indian Affairs, *Summary under the Criteria and Evidence*, 25. A tyee is a local headman, or community leader.

38 Buerge, *Chief Seattle*, 135.

39 Buerge, *Chief Seattle*, 136–42.

40 "Treaty of Medicine Creek, 1854," HistoryLink, February 20, 2003, www.historylink.org/File/5253; Buerge, *Chief Seattle*, 133–34.

41 Eckrom, *Remembered Drums*, 6–12.

42 Mary Isely, *Uncommon Controversy: Fishing Rights of the Muckleshoot, Puyallup, and Nisqually Indians; A Report Prepared for the American Friends Service Committee* (Seattle: University of Washington Press, 1970); Buerge, *Chief Seattle*, 134.

43 Buerge, *Chief Seattle*, 135; "Treaty of Point Elliott," Duwamish Tribe, accessed April 2, 2019, www.duwamishtribe.org/treaty-of-point-elliott. The signers were Ts'huahnti, Now-a-chais, and Ha-seh-doo-an.

44 "Treaty of Point No Point, 1855," HistoryLink, January 15, 2004, www.historylink .org/File/5637; Charles M. Gates, "The Indian Treaty of Point No Point," *Pacific Northwest Quarterly*, April 1955, https://www.jstor.org/stable/40487129.

45 Buerge, *Chief Seattle*, 143.

46 A. Denny, *Pioneer Days*, 69; Isely, *Uncommon Controversy*, 17.

47 Cassandra Tate, "Gold in the Pacific Northwest," HistoryLink, December 6, 2004, www.historylink.org/File/7162.

48 Eckrom, *Remembered Drums*, 21. McAllister arrived with the Simmons party in 1844. See Thomas Prosch, "The Political Beginning of Washington Territory," *Quarterly of the Oregon Historical Society*, 6, no. 2 (June 1905): 147–58.

49 Eckrom, *Remembered Drums*, 22; James McAllister to Superintendent of Indian Affairs, Washington Territory, October 16, 1855, USGenNet, www.usgennet.org /usa/wa/state/jamesmcallister.html.

50 Eckrom, *Remembered Drums*, 23–26.

51 A. Denny, *Pioneer Days*, 69.

52 Cordelia Hawk Putvin, "About Indians," in Daughters of the Pioneers, *Stories of the Pioneers: True Stories from Members of the Daughters of the Pioneers of Washington*, (Seattle, WA: State Association of the Daughters of the Pioneers of Washington, 1986), 20–22, quoted in "George Washington Bush: Washington State Pioneer," http://start-wa.com/indian_war.html, accessed November 25, 2019.

53 John King, letter reprinted in E. Denny, *Blazing the Way*, 91–95; Bagley, *History of Seattle*, 165–70.

54 Isely, *Uncommon Controversy*, 39–42. Governor Stevens and the Pacific commander of the US Army, John Ellis Wool, fought bitterly over Stevens's militaristic stance against the tribes. Wool argued that the tribes needed protection from the settlers, not the other way around. Complaints from local legislators to Congress eventually led to Wool's removal, leaving Stevens free to wage war against the resisting members of the tribes.

55 Bureau of Indian Affairs, *Summary under the Criteria and Evidence*; Reprint of "sketch" by Rev. G. F. Whitworth in E. Denny, *Blazing the Way*, 374–77.

56 Buerge, *Chief Seattle*, 157–58.

57 Prosch, "A Chronological History of Seattle," 78; E. Denny, *Blazing the Way*, 360; Rasmussen/Nelson family records. Tribal family records suggest that Owhi was related to Quio-litza, the Duwamish daughter of Kanum and Tupt-Aleut, who was then living at Lake Fork with her white husband, Dr. R. M. Bigelow.

58 Buerge, *Chief Seattle*, 159.

59 E. Denny, *Blazing the Way*, 69; Frank Conklin, "The Moses Family," unpublished manuscript, Renton History Museum archives, n.d.

60 Clarence Bagley, *Early Days in Seattle*, vol. 1 (Seattle, WA: Argus Press, 1920), 107.

61 E. Denny, *Blazing the Way*, 90; Isely, *Uncommon Controversy*, 42–44; Bagley, *History of King County*, 1:181. A new reservation at Muckleshoot Prairie and expanded Puyallup and Nisqually reservations were formalized by an executive order amending the terms of the Treaty of Medicine Creek.

62 Isely, *Uncommon Controversy*, 161.

63 Thrush, *Native Seattle*, 234; "Indians Burned Out: Exodus of the Red Men from West Seattle," *Seattle Press-Times*, March 7, 1893.

64 "Ordinances of the Town of Seattle," *Seattle Weekly Gazette*, March 4, 1865, cited in Jennifer Ott, "Seattle Board of Trustees Passes Ordinance, Calling for Removal of Indians from the Town, on February 7, 1865," HistoryLink, December 7, 2014, www.historylink.org/File/10979.

65 Washington Territory, Superintendent of Indian Affairs, *Report of the Washington and Oregon Superintendent of Indian Affairs for the Year 1857* (Washington, DC: Government Printing Office, 1857), 15–18; Bagley, *History of King County*, 1:191.

66 Whitworth sketch in E. Denny, *Blazing the Way*, 376–77. Smithers's closest neighbors when he first arrived were the family of R. M. Bigelow and Quio-litza (Ann).

67 United States, Department of the Interior, Office of Indian Affairs, *Annual Report of the Commissioner of Indian Affairs for the Year 1865: Washington Superintendency*, by W. H. Waterman (Washington, DC: Government Printing Office, 1865), 70, http://digicoll.library.wisc.edu/cgi-bin/History/History-idx?id=History .AnnRep65a.

68 Office of Indian Affairs, *Annual Report of the Commissioner of Indian Affairs for the Year 1865*, 70.

69 David Wilma, "Seattle Pioneers Petition against a Reservation on the Black River for the Duwamish Tribe in 1866," HistoryLink, January 24, 2001, www.historylink .org/File/2955.

70 Richard Walker, "10 Things You Should Know about the Duwamish Tribe," *Indian Country Today* (Washington, DC), July 16, 2015, https://newsmaven.io/indian countrytoday/pages/about-indian-country-today-zzNHHkzlzEWFeg4RvK- fKvQ; Oscar Halpert, "Right to Be Recognized," *Renton Reporter*, August 16, 2006.

71 Rasmussen/Nelson family records; Julia Anne Allain, "Duwamish History in Duwamish Voices: Weaving Our Family Stories since Colonization," PhD diss., University of Victoria, 2014.

72 "Dr. Reuben Miles Bigelow," Ancestry.com, accessed March 20, 2019; McLemore, Descendants of Mary Kennum; James Rasmussen interviews.

73 "Abner Jefferson Tuttle," Ancestry.com, accessed March 20, 2019; Jim Tuttle, historical documents, Vashon Heritage Museum, Vashon, WA.

74 United States Census, July 25, 1871, King County, WA, accessed through Ancestry .com; Rasmussen/Nelson family records.

75 "Seattle and King County Milestones," HistoryLink, October 27, 2004, www.history link.org/File/7110.

76 United States Census, 1879, Seattle, King County, WA, accessed through Ancestry .com.

77 United States Census, 1880, King County, WA, accessed through Ancestry.com.

78 Rasmussen/Nelson family records.

79 Bagley, *History of King County, Washington*; T. T. Waterman, with Vi Hilbert, Jay Miller and Zalmai Zahir, *Puget Sound Geography* (Federal Way, WA: Lushootseed Press, 2001).

80 List and acreage of public domain allotments to tribal members, signed by Office of Indian Affairs Superintendent O. C. Upchurch and marked as Exhibit #14, in author's possession; 1919 Supplement to United States Compiled Statutes, Annotated, vol. 1 (St. Paul, MN: West, 1920), 76, 96; Homestead Certificate No. 5931, signed by President William M. McKinley, August 18, 1897, accessed through General Land Office Records website, Bureau of Land Management, Department of the Interior, https://glorecords.blm.gov, accessed April 3, 2019. The Homestead Act replaced the Donation Land Claims Act in 1862.

81 Bagley, *History of King County, Washington*,1:137–43.

82 Bagley, *History of King County, Washington*, 1:137–43.

83 Bagley, *History of King County, Washington*,1:137–43.

84 Rasmussen/Nelson family records; Henry Flueck, Superior Court of the State of Washington, petition for letters of administration in the matter of the estate of Dr. Jack Bigelow, King County, WA; Judge Ronan, Superior Court of the State of Washington, June 5, 1910, decree of settlement and distribution in the matter of the estate of Dr. Jack Bigelow: No. 7729, King County, WA; "'New Black Diamond Mine' Employed 500 Workers, Produced 4,000 Tons of Coal in Two 8-Hour Shifts," Black Diamond History, November 29, 2017. https://black diamondhistory.wordpress.com/2017/11/29/new-black-diamond-mine-employed -500-workers-produced-4000-tons-of-coal-in-two-8-hour-shifts.

85 Rasmussen/Nelson family records.

86 School Land Contract, State of Washington, to M. A. Overacker, filed by King County Auditor, July 7, 1906, King County Archives; Rasmussen/Nelson family records.

CHAPTER 3: SWEAT AND THE TRANSFORMATION OF A WATERSHED

1 Valarie Bunn, "Seattle's Pioneers of Fremont: John Ross," Wedgwood in Seattle's History, February 8, 2016, https://wedgwoodinseattlehistory.com/2016/02/08 /seattles-pioneers-of-fremont-john-ross.

2 Ida Ross, "Memoirs of a Daughter, the Third Child of Mr and Mrs John Ross, Founders of the Ross School, the First School in the Ballard/Fremont Area," 1938, History Collection, Fremont Public Library, Seattle, WA.

3 Ross, "Memoirs of a Daughter," 10–12.

4 Ross, "Memoirs of a Daughter," 12.

5 John Ross v. The Washington Improvement Company, Frontier Justice Files #KNG-3398 (Third Judicial Court of Washington Territory [King County Superior Court] August 6, 1883).

6 Ross, "Memoirs of a Daughter," 12–13.

7 Washington Improvement Company v. John Ross, Frontier Justice Files #KNG-3737 (Third Judicial District Court of Washington Territory [King County Superior Court] March 5, 1984).

8 Washington Improvement Company v. John Ross; Ralph Johnson, "Memory Digs a Canal: The Creek," *Sea Chest*, June 8, 1975; David B. Williams and Jennifer Ott, *Waterway: The Story of Seattle's Locks and Ship Canal* (Seattle: HistoryLink, 2017).

9 Williams and Ott, *Waterway*.

10 Williams and Ott, *Waterway*; Clarence Bagley, *History of King County, Washington*, vol. 1 (Chicago: S. J. Clarke, 1929), 84.

11 Williams and Ott, *Waterway*.

12 Williams and Ott, *Waterway*.

13 Williams and Ott, *Waterway*, 49.

14 Eugene Semple to Adria Semple, April 27, 1897, Eugene Semple Papers, University of Washington Libraries, Special Collections, Seattle, WA.

15 Bagley, *History of King County*, 1:380–81; "Histories of the Canal Controversy," Eugene Semple Papers, Box 18, Folder 1, University of Washington Libraries, Special Collections, Seattle, WA.

16 David B. Williams, *Too High and Too Steep: Reshaping Seattle's Topography* (Seattle: University of Washington Press, 2015), 94–96.

17 Clarence B. Bagley, *History of Seattle: From the Earliest Settlement to the Present Time* (Chicago: S. J. Clarke, 1916), 380–89.

18 History [of the] South Canal and Harbor Improvements, publication, Seattle and Lake Washington Waterway Company (1902), Eugene Semple Papers.

19 Excavation of Waterways by Private Contract, Chapter XCIX, Session Laws (1893), Washington State Legislature, http://leg.wa.gov/CodeReviser/documents/session law/1893c99.pdf.

20 Bagley, *History of Seattle*, 380–81; *History [of the] South Canal and Harbor Improvements*, Seattle and Lake Washington Waterway Company, 1902, Eugene Semple Papers.

21 Robert C. Nesbit, *He Built Seattle: A Biography of Judge Thomas Burke* (Seattle, WA: University of Washington Press, 1961), 404; John J. McGilvra, "To Lower the Lake: No Relief from Private Canals, Judge McGilvra Thinks," *Seattle Post-Intelligencer*, July 15, 1894.

22 Bunn, "Seattle's Pioneers of Fremont: John Ross."

23 *The Ship Canal Connecting Lake Washington and Puget Sound: Its Advantages and the Reasons Why It Should be Constructed by the United States Government* (Seattle: Seattle Chamber of Commerce, n.d.).

24 Eugene Semple, "Chronology of Seattle Ship Canal," August 24, 1903, Eugene Semple Papers; Bagley, *History of Seattle*, 380–89; Eugene Semple to Adria Semple, April 27, 1897, Eugene Semple Papers.

25 Semple, "Chronology of Seattle Ship Canal," 4.

26 Eugene Semple, "Chronology of Seattle Ship Canal"; Eugene Semple to Colonel Thos W. Symons, June 27, 1905, Eugene Semple Papers; David Williams, "South

Canal Seattle Photo," GeologyWriter.com, January 28, 2015, http://geologywriter
.com/blog/south-canal-seattle-photo.

27 Williams and Ott, *Waterway.*

28 Semple to Symons, June 27, 1905.

29 Richard C. Berner, *Seattle 1900–1920: From Boomtown, Urban Turbulence, to Restoration* (Seattle: Charles Press, 1991), 17–20.

30 Semple, "Chronology of Seattle Ship Canal"; Williams and Ott, *Waterway.*

31 Semple, "Chronology of Seattle Ship Canal." Using a canal to drain the floodwaters
of Lake Washington and the Cedar River was first proposed by the King County
civil engineer, O. F. Wegener, in 1892 ("To Prevent Floods: Lake Washington
Canal Will Drain a Great Basin," *Seattle Post-Intelligencer*, November 14, 1892).
However, neither Semple's canal—which may have been inspired by Wegener's
proposal—nor the north canal backers had originally designed their projects for
this purpose.

32 Williams and Ott, *Waterway.*

33 Berner, *Seattle 1900–1920;* Williams, "South Canal Seattle Photo;" "Havoc Wrought
in Sluicing for Alleged South Canal," *Seattle Times*, May 8, 1904.

34 School Land Contract, State of Washington, July 7, 1906, King County Regional
Archives, Seattle.

35 School Land Contract; School Land Deed, State of Washington, March 19, 1907,
King County Regional Archives, Seattle.

36 Rasmussen/Nelson family records, genealogical, legal, and property records provided
by descendants of Kanum family, 1800s to present, in author's possession; James
Rasmussen, interviews with author, 2015–19.

37 Semple, "Chronology of Seattle Ship Canal," 21–22.

38 *Report of the Governor of Washington Territory to the Secretary of the Interior, 1887*
(Washington, DC: Government Printing Office), 52–54.

39 Semple, "Chronology of Seattle Ship Canal," 21–22; Semple to Symons, June 27, 1905.

40 Semple to Symons, June 27, 1905; King County Probate Records, Semple Estate
(No. 9463), 1908, Eugene Semple Papers.

41 Bagley, *History of Seattle,* 93–94.

42 John Dreher, "King County is Benefited by Flood," *Seattle Daily Times*, November 16, 1906; Lucile McDonald, "The White: The River Nobody Wanted," *Seattle
Daily Times*, January 27, 1957; William Hempel census records, 1900–1930,
accessed via Ancestry.com.

43 "Laid Waste by Water," *Seattle Post-Intelligencer*, November 21, 1892; "A Plague of
Water," *Seattle Post-Intelligencer*, May 21, 1893.

44 "To Prevent Floods."

45 "Valley Suffers by Disastrous Flood," *White River Journal* (Kent, WA), November 16,
1906; Alan J. Stein, "Flood Submerges South King County Beginning on November 14, 1906," HistoryLink, September 23, 2001, www.historylink.org/File/3585;
Alan J. Stein, "White River Valley (King County)—Thumbnail History," History-
Link, September 23, 2001, www.historylink.org/File/3583.

46 Stein, "Flood Submerges South King County"; McDonald, "The White."

47 Stein, "Flood Submerges South King County"; "Valley Suffers."

48 McDonald, "The White"; Stein, "Flood Submerges South King County"; King County, Department of Natural Resources and Parks, *Habitat Limiting Factors and Reconnaissance Assessment Report: Green/Duwamish and Central Puget Sound Watersheds—Executive Summary*, December 2000, https://your.kingcounty.gov /dnrp/library/2000/kcr728/execsum/execsum.pdf.

49 Cynthia Updegrave, "Cedar-Sammamish Watershed," HistoryLink, December 20, 2016, www.historylink.org/File/20273.

50 David Williams, "Olmsted Parks in Seattle," HistoryLink, May 10, 1999, www.history link.org/File/1124.

51 Mimi Sheridan and Carol Tobin, "Seattle Neighborhoods: Licton Springs: Thumb-nail History," HistoryLink, July 17, 2001, www.historylink.org/File/3447; Wil-liams, "Olmsted Parks in Seattle"; Jason King, "Seattle: Green Lake," Hidden Hydrology, October 26, 2018, www.hiddenhydrology.org/seattle-green-lake.

52 Updegrave, "Cedar-Sammamish Watershed"; Morda C. Slauson, *One Hundred Years along the Cedar River* (Renton, WA: Maple Valley Historical Society, 1971), 3.

53 David M. Buerge, "Requiem for a River," *Seattle Weekly*, October 16, 1985.

54 King County, Department of Natural Resources and Parks, *Habitat Limiting Factors and Reconnaissance Assessment Report*. Had Semple's canal been constructed instead of the northern route, Lake Washington would not have been lowered, and the Black River might still flow today. The watershed would be intact, albeit with an additional outlet to Elliott Bay. It is difficult to imagine the difference that decision might have made in the history and fortunes of the watershed, its people, and the city itself.

55 David Wilma, "Harbor Island, at the Time the World's Largest Artificial Island, Is Completed in 1909," HistoryLink, November 6, 2001, www.historylink.org/File /3631; H. W. McCurdy Collection on the Puget Sound Bridge & Dredging Com-pany, 1900–1945, Box 1, Folder 3, Museum of History and Industry, Seattle, WA.

56 Bagley, *History of Seattle*, 628; Norman Reed, "Flour Milling in Washington: A Brief History," HistoryLink, July 11, 2010, www.historylink.org/File/9474.

57 Thomas W. Prosch, "A Chronological History of Seattle from 1850 to 1897, Prepared in 1900 and 1901," unpublished manuscript, University of Washington Libraries, Northwest Collection, Seattle, WA, 287; Bagley, *History of Seattle*, 628. The Morans opened their shipyard under the name Seattle Dry Dock and Shipbuilding, chang-ing the name to Moran Brothers Company in 1889.

58 Rasmussen interviews.

59 Virgil Gay Bogue, *Plan of Seattle: Report of the Municipal Plans Commission* (Seattle, WA: Lowman and Hartford, 1911); "To Prevent Floods"; *The Duwamish Diary, 1849–1949* (Seattle, WA: Cleveland High School, Seattle Public Schools, 1949), 65–72; Hiram M. Chittenden, *Report of an Investigation by a Board of Engineers of the Means of Controlling Floods in the Duwamish-Puyallup Valleys and Their Tributaries in the State of Washington*, United States, Army Corps of Engineers,

Joint Committee of King and Pierce Counties on Flood Situation in Duwamish and Puyallup Valleys, Washington, (Seattle, WA: Lowman and Hanford, 1907), 21–22.

60 Elsa Bowman, personal communication, February 11, 2019; Richard Hamm and Hamm family descendants, interview with author, October 31, 2018; Paul Dorpat, "Seattle Now and Then: The Butler Did It!," Seattle Now and Then, https://paul dorpat.com/2017/01/21/seattle-now-then-the-butler-did-it, accessed December 2, 2019.

61 Dorpat, "Seattle Now and Then."

62 Richard Hamm and Hamm family descendants interview.

63 Lewis Hamm, "Hamm's Creek," 1995, memoir, in author's possession.

64 Bogue, Plan of Seattle. Orillia was a community between present-day Renton and Kent.

65 Paul Dorpat, "The Hamm Home," Seattle Times, n.d.; Richard Hamm and Hamm family descendants interview. The value of bonds paid by Duwamish Valley landowners totaled more than $1.8 million, more than $100,000 of which would be assessed to Hamm's estate. See Duwamish Diary.

66 Richard Hamm and Hamm family descendants interview.

67 Bowman, personal communication, February 11, 2019.

68 Fred W. Newell, Frank Powell, and John Shorett, "In Memoriam," eulogy by commissioners and coworkers of Duwamish Waterway, 1918, in author's possession; John B. Shorett to Charlie Hamm, October 2, 1918, in author's possession.

69 Bogue, Plan of Seattle; Bagley, History of Seattle, 357.

70 "Launch Sound's Only Public Owned Dredge," Seattle Daily Times, May 22, 1913; "New Dredge Given Successful Trials," Seattle Daily Times, September 7, 1913; "New Waterway District Dredge Begins Operation," Seattle Daily Times, September 17, 1913.

71 "County's Title to Land in Duwamish Valley Confirmed," Seattle Daily Times, March 16, 1913; "Big Dredge Gouges Out Turning Basin," Seattle Daily Times, November 9, 1913.

72 "Celebrate Beginning of Duwamish Project," Seattle Daily Times, October 14, 1913; "Duwamish Project Launched," Seattle Daily Times, October 16, 1913.

73 Newell v. Loeb, 137 P.811 (Wash. 1913).

74 "White City to Be Built at Oxbow," Seattle Daily Times, October 25, 1906; "Realty Activity Grows More and More Pronounced Every Week," Seattle Daily Times, October 28, 1906; Newell v. Loeb, 137 P.811 (Wash. 1913).

75 Newell v. Loeb, 137 P.811 (Wash. 1913).

76 Newell v. Loeb, 137 P.811 (Wash. 1913).

77 "South End to See Awakening," Seattle Daily Times, January 3, 1915; "Part of Duwamish Waterway Opened," Seattle Daily Times, March 28, 1915.

78 "Big Dredge Disabled," January 14, 1916, newspaper clipping in author's possession; "Work on Oxbow Bend Fill Is Progressing Slowly," February 18, 1916, newspaper clipping in author's possession.

79 "Georgetown Is Growing Fast: Many Industries Moving In," *Seattle Daily Times*, January 3, 1926.

80 Peter Blecha, "Rainier Beer: Seattle's Iconic Brewery," HistoryLink, August 26, 2009, www.historylink.org/File/9130; Gary Flynn, "Albert Braun Brewing Association (1891–1893)," Brewery Gems, www.brewerygems.com/braun.htm, accessed April 10, 2019; *Historic Duwamish News*, Greater Duwamish Historic Preservation Society, n.d., in author's possession.

81 Thomas Veith, *History of South Park* (Seattle: City of Seattle, Department of Neighborhoods, Historic Preservation Program, 2009), 43–44; Bagley, *History of Seattle*, 623–26.

82 "The Great Seattle Fire," University Libraries, University of Washington, n.d., accessed April 10, 2019, https://content.lib.washington.edu/extras/seattle-fire.html.

83 "The Great Seattle Fire."

84 Bagley, *History of Seattle*, 623–24; "Pillars of Business Structure," *Seattle Daily Times*, February 22, 1927; "Urging Duwamish Improvement," *Seattle Daily Times*, December 6, 1903.

85 Articles of Incorporation of the Puget Sound Fire Clay Company, August 23, 1889, Puget Sound Regional Archives, Bellevue, WA; Alan Stein, "Denny, Orion O. (1853–1916)," HistoryLink, November 24, 2002, www.historylink.org/File/4026; "Ghost Town of Taylor, King County, Washington," Ghost Towns of Washington, www.ghosttownsofwashington.com/taylor.html, accessed April 10, 2019; "Washington Bricks," Brickmakers, http://washingtonbricks.com/brick.dennyclayco taylorbm.html, accessed April 10, 2019.

86 Veith, *History of South Park*, 46–49; Blecha, "Rainier Beer"; Walt Crowley, "Georgetown Votes to Annex to Seattle on March 29, 1910," HistoryLink, March 28, 2000, www.historylink.org/File/2663.

87 Bagley, *History of Seattle*, 602–5; "Albert Premel," Ancestry.com, www.ancestry.com /family-tree/person/tree/9202019/person/-670199098/facts, accessed April 10, 2019; Marianne Clark, interview with author, December 19, 2015. The horse stables where Marianne Clark's grandfather likely worked still stands, just off present-day Airport Way. Her house sits on the Van Asselt land claim, sold to Vulcan Iron Works in 1902. See Emily Inez Denny, *Blazing the Way: Or, True Stories, Songs and Sketches of Puget Sound and Other Pioneers (1899)* (Whitefish, MT: Kessinger, [1909] 2011), 321–22; Marianne's granddaughter is the sixth generation of her family to live in the Duwamish Valley.

88 "Duwamish Valley: The Men Who Have Made It," *Seattle Star*, January 20, 1910; Ryan M. Thompson and Chris Carter, *Terminal 115 Environmental Conditions Report*, Port of Seattle (Seattle: Sound Earth Strategies, 2011), 12; Paul Spitzer, "Harsh Ways: Edward W. Heath and the Shipbuilding Trade," *Pacific Northwest Quarterly* 90, no. 1 (1998): 3–16, www.jstor.org/stable/40492441.

89 Walt Crowley, "Boeing and Early Aviation in Seattle, 1909–1919," HistoryLink, March 5, 2003, www.historylink.org/File/5369.

90 Crowley, "Boeing and Early Aviation."

91 Thompson and Carter, *Terminal 115 Environmental Conditions Report*, 20, figures 4B, 4C.

92 Walt Crowley, "Turning Point 6," HistoryLink, April 5, 2001, www.historylink.org /File/9300; "Boeing Model 247," Aviation History Online Museum, www.aviation -history.com/boeing/247.html, accessed August 8, 2019.

93 Rita Creighton, "The Early History of KCIA/Boeing Field," Airport Journals, May 1, 2005, http://airportjournals.com/the-early-history-of-kciaboeing-field.

94 "The Duwamish, Industrial Artery," *Seattle Times*, November 25, 1956; Priscilla Long, "South Park Bridge, Duwamish Waterway, King County, 1931–2010," His-toryLink, October 17, 2014, www.historylink.org/File/10937; Suzanne Hittman, interview with author, September 29, 2018; Rita Cipalla, "Joe Desimone: From Produce Farmer to Owner of Seattle's Pike Place Market," HistoryLink, April 1, 2018, www.historylink.org/File/20529.

95 Pike Place Market Records, 1894–1990, Archives West, http://archiveswest.orbis cascade.org/ark:/80444/xv26684; Hittman interview.

96 "Hoovervilles in Seattle," n.d., Seattle Municipal Archives, www.seattle.gov /cityarchives/exhibits-and-education/digital-document-libraries/hoovervilles -in-seattle; "Skinner and Eddy Shipyard, Plant #2, Seattle, WA," PCAD (Pacific Coast Architecture Database), http://pcad.lib.washington.edu/building/17253, accessed March 28, 2019.

97 Hittman interview; David Sweeney, interview with author, October 1, 2018.

98 Priscilla Long, "South Park Bridge, Duwamish Waterway, King County, 1931–2010," HistoryLink, October 17, 2014, www.historylink.org/File/10937; Tom Reese and Eric J. Wagner, *Once and Future River: Reclaiming the Duwamish* (Seattle: Univer-sity of Washington Press, 2016), 27–34.

99 Lucile Carlson, "Duwamish River: Its Place in the Seattle Industrial Plan," *Eco-nomic Geography*, 26, no. 2 (April 1950): 148, www.jstor.org/stable/141732; Lars Langloe, *Report on Development of Industrial Sites in the Duwamish-Green River Valley*, (Seattle: City of Seattle, City Planning Commission, 1946), 1–2.

100 Langloe, *Report on Development*, 1–2, 4, 8, 20.

101 Carlson, "Duwamish River," 50, 153.

102 "Howard Hanson Dam: Background," Seattle District, US Army Corps of Engi-neers, www.nws.usace.army.mil/Missions/Civil-Works/Locks-and-Dams/Howard -Hanson-Dam; Alan Stein, "Howard A. Hanson Dam," HistoryLink, Septem-ber 10, 2001, www.historylink.org/File/3549.

103 Devon Stark, "A Brief History of Salmon Fishing in the Pacific Northwest," Mid Sound Fisheries Enhancement Group, n.d., www.midsoundfisheries.org/a-brief -history-of-salmon-fishing-in-the-pacific-northwest, accessed April 10, 2019; Joseph E. Taylor, *Making Salmon: An Environmental History of the Northwest Fisheries Crisis* (Seattle: University of Washington Press, 2009).

104 Robert Higgs, "Legally Induced Technical Regress in the Washington Salmon Fishery," Independent Institute, June 30, 1982, www.independent.org/publications /article.asp?id=2453#3, accessed April 10, 2019.

105 Matthew W. Klingle, *Emerald City: An Environmental History of Seattle* (New Haven, CT: Yale University Press, 2008), 174–75; Taylor, *Making Salmon*; Tom Dailey, "Coast Salish Villages of Puget Sound," http://coastsalishmap.org/start _page.htm (see map 6, site 44), accessed April 10, 2019. "Trap sites" were licensed starting in 1898, including many at locations where Native tribes had been catching salmon for millennia.

106 Klingle, *Emerald City*.

107 Klingle, *Emerald City*, 177, citing the Duwamish, Lummi, Whidbey Island, Skagit, Upper Skagit, Swinomish, Kikiallus, Snohomish, Snoqualmie, Stillaguamish, Suquamish, Samish, Puyallup, Squaxin, Skokomish, Upper Chehalis, Muckleshoot, Nooksack, Chinook, and San Juan Islands Tribes of Indians, Claimants, v. the United States of America, Defendant, Consolidated petition No. F-275, 2 vols. (Seattle: Argus, ca. 1933); "Brailing a Puget Sound Salmon Trap: 60,000 Sockeye Were Taken by This Trap in One Catch," 1910 photograph, University Libraries, University of Washington Digital Collections, https://digitalcollections .lib.washington.edu/digital/collection/fishimages/id/44504/rec/16.

108 Higgs, "Legally Induced Technical Regress."

109 Higgs, "Legally Induced Technical Regress," 3–8, citing Washington State, Sessions Laws, 1938; "Lethal Fish Trap Gets Second Chance as Recovery Tool," Wild Fish Conservancy Northwest, http://wildfishconservancy.org/resources/publications /wild-fish-runs/lethal-fish-trap-gets-second-chance, accessed April 11, 2019.

110 Stein, "Howard A. Hanson Dam."

111 Shannon Sawyer, "Lower Green River Valley Agricultural Production District (ADP) Is One of Five King County ADPs Designated on April 8, 1985," History Link, December 18, 2018, www.historylink.org/File/20697.

112 David Takami, "World War II Japanese American Internment: Seattle/King County," HistoryLink, November 6, 1998, www.historylink.org/File/240; "About the Incarceration," Densho Encyclopedia, n.d., http://encyclopedia.densho.org /history, accessed March 26, 2019.

113 Sharon Bowell and Lorraine McConaghy, "Abundant Dreams Diverted," *Seattle Times*, June 23, 1996, http://old.seattletimes.com/special/centennial/june/intern ment.html; Hittman interview.

114 Hittman interview.

115 Amanda Zahler, unpublished research for Amanda Zahler, Anna Marti, and Gary Thomsen, *Seattle's South Park: Images of America Series* (Mount Pleasant, SC: Arcadia Publishing, 2006), provided by Anna Marti; Hittman interview.

116 Rasmussen interviews, 2015–19.

117 David Yamaguchi, interview with author, August 6, 2018.

118 Yamaguchi interview; Catherine Natsuko Yamaguchi to the Overacker family, May 24, 1942, in Rasmussen/Nelson family records. Natsuko was photographed by Ansel Adams at Manzanar during the war: see Catherine Natsuko Yamaguchi, Red Cross Instructor, Library of Congress, www.loc.gov/item/2001704616.

119 Brian F. Atwater, Satoko Musumi-Rokkaku, Kenji Satake, Yoshinobu Tsuji, Kazue Ueda, and David K. Yamaguchi, *The Orphan Tsunami of 1700: Japanese Clues to a Parent Earthquake in North America* (Seattle: US Geological Survey, 2005), https://pubs.er.usgs.gov/publication/pp1707; Ruth S. Ludwin, Gregory Smits, Deborah Carver, et al., "Folklore and Earthquakes: Native American Oral Traditions from Cascadia Compared with Written Traditions from Japan," *Myth and Geology* 273 (January 2007): 67–94.

120 Zahler, Marti, and Thomsen, *Seattle's South Park: Images of America Series* (Mount Pleasant, SC: Arcadia Publishing, 2006); Rosario Maria-Medina Barron, interview with author, April 4, 2018; Wally Barron, interview with author, July 29, 2018.

CHAPTER 4: TEARS ON THE FENCELINE

1 Byron Johnsrud, "Embattled Housewives Picket Dump," *Seattle Daily Times*, November 1, 1961.

2 Rev. Robert Morton, "Council Should Act Immediately on South Park Dump," *Seattle Daily Times*, November 7, 1961.

3 "South End Fire Dump Protested," *Seattle Daily Times*, March 2, 1959; "Burning Dump," *Seattle Daily Times*, April 1, 1959; "Huge Incinerator, Monthly Garbage Charges Proposed," *Seattle Daily Times*, August 25, 1960; Amanda Zahler, Anna Marti, and Gary Thomsen, *Seattle's South Park: Images of America Series* (Mount Pleasant, SC: Arcadia Publishing, 2006).

4 Douglas Willix, "South Park Refuse Fires to Die," *Seattle Daily Times*, November 29, 1961; "When Women Speak Out in Protest," *Seattle Daily Times*, November 30, 1961.

5 City of Seattle, Seattle Planning Commission, 1956 Seattle Comprehensive Plan, Resolution No. 17488, April 20, 1957, A Resolution Adopting in Principle the Comprehensive Plan for the City of Seattle, Seattle Municipal Archives, Seattle, WA.

6 Douglas Willix, "South Park Zoning Hearing Recessed," *Seattle Daily Times*, January 11, 1967; Anna Marti, "South Park Timeline," unpublished description of significant events, in author's possession.

7 Jackie Jacquemart, interview with author, October 22, 2018; Michael Sweeney, "South Park: A Square Mile of Defiance," *Seattle Post-Intelligencer*, August 11, 1974; "South Park Residents Score in Battle against Industry," *Seattle Daily Times*, February 27, 1968.

8 Walt Woodward, "South Park Situation Dramatizes the Dilemma of Zoning," *Seattle Daily Times*, March 3, 1968.

9 Greg Wingard, interview with author, September 13, 2018.

10 Wingard interview.

11 Wingard interview.

12 Wingard interview; David Suffia, "Dump at Old Mine May Endanger Kent, Seattle Water," *Seattle Times*, September 16, 1983.

13 Paul Roberts, "A Mine is a Terrible Place for Waste," *Seattle Weekly*, April 10, 1991; Wingard interview, September 13, 2018; Lee Dorrigan, interview with author, July 27, 2018.

14 Wingard interview; Facility Site ID No. 2139, Toxic Waste Cleanup Division case file, Washington State Department of Ecology, Bellevue, WA.

15 Dorrigan interview.

16 Clarence Bagley, *History of King County, Washington,* vol. 1 (Chicago: S. J. Clarke, 1929), 280–300; Laura McCarty, "Coal in the Puget Sound Region," HistoryLink, January 31, 2003, www.historylink.org/File/5158; Harry Stuart Hutchinson, "Coal in the State of Washington, 1950," master's thesis, College of Puget Sound, 18–36; Danny Westneat, "Wilson: Boeing Followed the Rules," *Valley Daily News* (Kent, WA), August 31, 1990; Danny Westneat, "13 Companies Dumped Toxins in Fallen Mine," *Valley Daily News* (Kent, WA), May 9, 1991.

17 Mary Ann Gwinn, "Boeing Waste Disposal at Issue: Federal Court Trial May Yield Answers," *Seattle Times*, August, 22, 1990; "Queen City Farms Site Profile," Environmental Protection Agency, October 20, 2017, https://cumulis.epa.gov/supercpad /SiteProfiles/index.cfm?fuseaction=second.Cleanup&id=1000835#bkground.

18 Mary Ann Gwinn, "It's 'Boeing Bashing,' Says Firm's Lawyer: Pollution Cleanup Trial Goes to Jury," *Seattle Times*, September 19, 1990; Ken Jensen, "Landsburg Mine: Little-Known Miners' Memorial Is the Site of Tragedies in 1954 and 1955," Black Diamond History, January 7, 2014, https://blackdiamondhistory.wordpress .com/2014/01/07/landsburg-mine-little-known-miners-memorial-is-the-site-of -tragedies-in-1954-and-1955.

19 Betty (Morris) Falk and William Kombol, "Morris Brothers Coal Mining Company, Inc.," HistoryLink, December 12, 2007, www.historylink.org/File/8420; Nina Elizabeth (Morris) Falk and William Kombol, "Durham: A King County Coal Mining Town," HistoryLink, November 1, 2006, www.historylink.org/File/7996; William Kombol, interview with author, August 8, 2018.

20 Danny Westneat, "Drivers Lined Up to Get into Dump," *Valley Daily News* (Kent, WA), May 9, 1991; Kombol interview.

21 Farallon Consulting et al., *Data Summary Report: West of 4th Groundwater Investigation Area,* January 22, 2008, 3–1, www.farallonconsulting.com/sites/default/files /Data%20Summ %20rpt%20with%20att.pdf; Dorrigan interview.

22 Danny Westneat, "Years of Toxic Dumping Linger for Truck Driver," *Valley Daily News* (Kent, WA), August 19, 1990; Westneat, "Drivers Lined Up."

23 Westneat, "Years of Toxic Dumping"; Westneat, "13 Companies Dumped Toxins."

24 Dorrigan interview; Golder Associates, *Final Remedial Investigation/Feasibility Study (RI/FS) for the Landsburg Mine Site,* Washington State Department of Ecology, February 1, 1996; Washington Department of Ecology, *Landsburg Mine Cleanup Site (King County, Washington State),* video, YouTube, October 24, 2013, www.youtube.com/watch?v=fq0OhSSXPbo.

25 Golder Associations, *Final Remedial Investigation/Feasibility Study;* Greg Wingard, personal communication, October 12, 2018.

26 Wingard, personal communication, October 12, 2018.

27 Kombol interview; Wingard interview.

28 "33 U.S. Code § 407: Deposit of Refuse in Navigable Waters Generally," Legal Information Institute, www.law.cornell.edu/uscode/text/33/407.

29 Robert A. Shanley, "Franklin D. Roosevelt and Water Pollution Control Policy," *Presidential Studies Quarterly* 18, no. 2 (Spring 1988): 319–30; "Federal Water Pollution Control Act (1948)," Encyclopedia.com, www.encyclopedia.com/history /encyclopedias-almanacs-transcripts-and-maps/federal-water-pollution-control -act-1948.

30 Jennifer Latson, "The Burning River That Sparked a Revolution," *Time*, June 22, 2015; Elliott Bay Trustee Council, *Pre-assessment Screen: Lower Duwamish River*, December 2, 2009.

31 "Introduction to the Clean Water Act," United States Environmental Protection Agency, https://cfpub.epa.gov/watertrain/moduleFrame.cfm?parent_object_id=1996.

32 "33 U.S. Code § 1365: Citizen Suits," Legal Information Institute, www.law.cornell .edu/uscode/text/33/1365.

33 Matthew Klingle, *Emerald City: An Environmental History of Seattle* (New Haven, CT: Yale University Press, 2008), 203–29.

34 Klingle, *Emerald City*, 222.

35 Klingle, *Emerald City*, 220.

36 Klingle, *Emerald City*, 225–26.

37 Klingle, *Emerald City*, 225–26; Joseph E. Taylor, *Making Salmon: An Environmental History of the Northwest Fisheries Crisis* (Seattle: University of Washington Press, 2009).

38 Klingle, *Emerald City*, 226.

39 Klingle, *Emerald City*, 227; State v. McCoy, 63 Wn.2d 421 (1963), Justia Law, https: //law.justia.com/cases/washington/supreme-court/1963/36224-1.html; Gabriel Chrisman, "The Fish in Protests at Frank's Landing," Seattle Civil Rights and Labor History Project, 2008, https://depts.washington.edu/civilr/fish-ins.htm, accessed April 12, 2019.

40 Cecile Hansen, interview with author, July 27, 2017; "Bernice White Site Blessed," *Muckleshoot Monthly*, March 15, 2013. Manny Oliver and his siblings are directly descended from one of Se'alth's brothers. Manny passed away in 1998 while fishing in Tulalip Bay.

41 "Boldt Decision and Native American Treaty Rights," C-SPAN, January 13, 2014; Walt Crowley and David Wilma, "Federal Judge George Boldt Issues Historic Ruling Affirming Native American Treaty Fishing Rights on February 12, 1974," HistoryLink, February 23, 2003, www.historylink.org/File/5282; Dee Norton, "Federal Suit Attacks Indian Fishing Laws," *Seattle Daily Times*, September 19, 1970.

42 Crowley and Wilma, "Federal Judge George Boldt Issues Historic Ruling."

43 Robert O. Marritz, "Frank Found Vindication with Historic Boldt Decision," *Nisqually Valley News* (Yelm, WA), August 4, 2016, www.yelmonline.com/article_e9469868 -5a69-11e6-ad6d-278db7b63b7a.html; "Salmon and Steelhead Co-management,"

Washington Department of Fish and Wildlife, https://wdfw.wa.gov/fishing
/tribal/co-management; "The Unintended Consequences of the Boldt Decision,"
Cultural Survival, June 1987, www.culturalsurvival.org/publications/cultural
-survival-quarterly/unintended-consequences-boldt-decision; United States v.
State of Washington, 384 F. Supp. 312 (W.D. Wash. 1974), Justia Law, https://law
.justia.com/cases/federal/district-courts/FSupp/384/312/1370661.

44 David Schaefer, "Boldt's Last Ruling Stands, for Now," *Seattle Times*, January 25,
1995.

45 United States v. State of Washington, 476 F. Supp. 1101 (W.D. Wash. 1979), Justia
Law, https://law.justia.com/cases/federal/district-courts/FSupp/476/1101/1378614;
Duwamish Indian Tribe; Snohomish Indian Tribe; Steilacoom Indian Tribe,
Plaintiffs-Intervenors-Appellants, v. State of Washington, No. 95-35202, Findlaw,
October 23, 1996, https://caselaw.findlaw.com/us-9th-circuit/1413758.html.

46 Muckleshoot Indian Tribe, www.muckleshoot.nsn.us, accessed April 12, 2019.

47 "Bernice White Site Blessed."

48 Hansen interview; Peter Blecha, "Hansen, Cecile: Tribal Chairwoman of Seattle's
Duwamish People," HistoryLink, March 21, 2009, www.historylink.org/File/8963.

49 Jennifer Ott, "Wastewater Treatment and the Duwamish River," HistoryLink,
July 14, 2016, www.historylink.org/File/11250.

50 Cari Simson, Maggie Milcarek, and Dan Klempner, *Duwamish Valley Vision
Map and Report*, ed. BJ Cummings (Seattle: Duwamish River Cleanup Coalition/
TAG, 2009).

51 Ott, "Wastewater Treatment"; *Duwamish River Superfund Fact Sheet*, 2001, Duwamish River Cleanup Coalition, Seattle, WA; Matthew Klingle, "Burdens of History Haunt the Duwamish," *Seattle Post-Intelligencer*, December 6, 2007.

52 SAIC, *Lower Duwamish Waterway, RM 0.9 to 1.0 East, Slip 1: Summary of Existing
Information and Identification of Data Gaps*, Washington State Department of
Ecology, August 2008; Raymond J. Eineigl, US Army Corps of Engineers, to
J. Eldon Opheim, Port of Seattle, September 29, 1975, Port of Seattle Public
Records, Seattle WA; Polly Lane, "For the Birds? Island, Bridge Intertwined
in Dispute," *Seattle Times*, February 15, 1976.

53 "Uncovering Seattle's Hidden History," *Seattle Post-Intelligencer*, February 12, 1977;
Scott Smith, "The Kellogg Island Controversy," unpublished manuscript, December 16, 1976, in author's possession.

54 Pacific Groundwater Group, *T-108 Groundwater and Shoreline Soil Investigation
Final Work Plan*, Port of Seattle, May 3, 2006, 2; SAIC, *Lower Duwamish
Waterway*.

55 John Cronin and Robert F. Kennedy Jr., *The Riverkeepers: Two Activists Fight to
Reclaim Our Environment as a Basic Human Right* (New York: Simon & Schuster,
1999); Gwendolyn Chambers, "A Brief History," Riverkeeper, June 16, 2009, www
.riverkeeper.org/riverkeeper-mission/our-story/a-brief-history.

56 Chambers, "A Brief History."

57 Cronin and Kennedy, *The Riverkeepers;* Chambers, "A Brief History."

58 Cronin and Kennedy, *The Riverkeepers*.

59 Cronin and Kennedy, *The Riverkeepers*.

60 Puget Soundkeeper Alliance v. Cadman (Seattle) Inc. and Tilbury Cement Company, Complaint (United States District Court, Western District of Washington, C95-0489, filed March 30, 1995); Puget Soundkeeper Alliance v. Cadman (Seattle) Inc. and Tilbury Cement Company, Consent Decree (United States District Court, Western District of Washington, C95-0489, filed May 22, 1995).

61 "Manufacturer to End Pollution," *Seattle Times*, September 20, 1996.

62 Stephen Miller, "John Beal (1950–2006)," HistoryLink, June 8, 2006, www.history link.org/File/20578; Liana Beal, interview with author, June 1, 2017.

63 Liana Beal interview; Miller, "John Beal." Agent Orange is a dioxin-based defoliant chemical suspected of causing cancer in people exposed to its wartime use in Vietnam.

64 Miller, "John Beal"; John Beal, personal communications, 1996–2006.

65 Miller, "John Beal."

66 John Beal, personal communications, 1996–2006.

67 "Hamm Creek Source Control Project," map, Your King County, https://your.king county.gov/dnrp/library/water-and-land/watersheds/green-duwamish/hamm -creek-aerial-map.pdf.

68 Dan Cargill, interview with author, June 9, 2017.

69 Liana Beal interview; John Beal, personal communications, 1996–2006. Salmon hatch in freshwater, migrate to the open ocean for two to five years (depending on the species), and return to their natal streams at the end of their lives to spawn.

70 James Rasmussen, interviews with author, 2015–19.

71 Priscilla McLemore, descendants of Mary Kennum, genealogical record of descendants of Mary Kennum aka Tyee Mary or Tupt-Icut, Duwamish Tribal Office, Seattle, WA; "Leadership," Duwamish Tribe, www.duwamishtribe.org/leader ship, accessed March 20, 2019; Rasmussen interviews; James Rasmussen, address to Duwamish Valley Youth Corps Meeting, Seattle, WA, February 14, 2019, www .facebook.com/duwamish.valley.youth/videos/1220298354813093. The Duwamish Tribe codified their modern constitution nine years before the federal Indian Reorganization Act sought to establish similar constitutions for all recognized tribal nations.

72 Rasmussen interviews.

73 John Beal to Lewis K. Hamm, February 8, 1995, in author's possession; "Hamm Creek, Rediscovery Point," Duwamish Alive Coalition, www.duwamishalive.org /duwamish-sites/hamm-creek-rediscovery-point, accessed March 20, 2019.

74 Linnea Westerlind, "Marra-Desimone Park," Year of Seattle Parks, January 12, 2010, www.yearofseattleparks.com/2010/01/12/marra-desimone-park.

75 Waste Action Project v. Holnam, Inc., Consent Decree, C95-1212D (United States District Court, Western District of Washington). It was later determined that the Lost Fork was a remnant of a creek that flowed north through South Park, rather than of the east-flowing Hamm Creek, but the extensive piping of the

neighborhood's springs and creeks commingled the original drainage until its origins were nearly untraceable.

76 Miller, "John Beal."

77 George Blomberg, Charles Simenstad, and Paul Hickey, "Changes in Duwamish River Estuary Habitat over the Past 125 Years," *First Annual Meeting on Puget Sound Research*, vol. 2, (Seattle: Puget Sound Water Quality Authority, 1988), 437–54, www.eopugetsound.org/sites/default/files/features/resources/PugetSound Research1988Vol2Optimized_0.pdf.

78 Richard Strickland and Randy Schuman, "Puget Sound Voices: Don Malins Interview," Encyclopedia of Puget Sound, May 2013, www.eopugetsound.org/articles /puget-sound-voices-don-malins-interview.

79 "National Priorities List (NPL) Sites: By Listing Date," United States Environmental Protection Agency, June 4, 2018, www.epa.gov/superfund/national -priorities-list-npl-sites-listing-date. *Superfund* is the informal name for the law, which was amended in 1986 and is officially known as the Comprehensive Environmental Response, Compensation, and Liability Act (CERCLA). Superfund sites are sites that have been added to the National Priorities List for ranking and cleanup.

80 James Meador, interview with author, January 8, 2019; James Meador, Tracy K. Collier, and John E. Stein, "Determination of a Tissue and Sediment Threshold for Tributyltin to Protect Prey Species of Juvenile Salmonids Listed under the U.S. Endangered Species Act," *Aquatic Conservation* 12, no. 5 (September–October 2002), 539–51.

81 Meador interview.

82 United States Environmental Protection Agency, Environmental Cleanup Office, *EPA Superfund Record of Decision: Harbor Island* (Seattle, WA, September 11, 2003); Usha Varanasi, Edmundo Casillas, Mary R. Arkoosh, Tom Hom, David A. Misitano, Donald W. Brown, Sin-Lam Chan, Tracy K. Collier, Bruce B. McCain, and John E. Stein, *Contaminant Exposure and Associated Biological Effects in Juvenile Chinook Salmon* (Oncorhynchus tshawytscha) *from Urban and Nonurban Estuaries of Puget Sound*, United States National Oceanic and Atmospheric Administration, National Marine Fisheries Service, April 1993; J. P. Meador, J. E. Stein, W. Reichert, and U. Varanasi, "Bioaccumulation of Polycyclic Aromatic Hydrocarbons by Marine Organisms," *Review of Environmental Contamination and Toxicology* 143 (February 1995): 79–165.

83 Bob Lane, "Duwamish Cleanup Hailed as a Success," *Seattle Times*, May 31, 1989.

84 Bill Dietrich, "NOAA Sues City, Metro over Pollution," *Seattle Times*, March 20, 1990.

85 United States of America, et al. v. The City of Seattle and Municipality of Metropolitan Seattle, Consent Decree C90-395WD (United States District Court, Western District of Washington, December 23, 1991); Eric Pryne, "Water Settlement a Drop in Bucket," *Seattle Times*, September 4, 1991.

1 Environmental Protection Agency, "National Priorities List for Uncontrolled Hazardous Waste Sites," *Federal Register* 66, no. 178 (September 13, 2001): 47583–92, www.govinfo.gov/content/pkg/FR-2001-09-13/pdf/01-22741.pdf.

2 Donald C. Malins, "What's Happening to Our Fish?" *NOAA Magazine*, March–April 1980, www.eopugetsound.org/sites/default/files/features/resources/malins_noaa-magazine_mar-apr-1980_opt.pdf.

3 "EPA Bans PCB Manufacture; Phases Out Uses," EPA news release, April 19, 1979, https://archive.epa.gov/epa/aboutepa/epa-bans-pcb-manufacture-phases-out-uses.html.

4 Roy F. Weston, Inc., *Site Inspection Report: Lower Duwamish River (RM 2.5 to 11.5), Seattle, Washington,* by (Seattle: United States Environmental Protection Agency, Region 10, 1999); Hal Bernton, "Group to Examine Duwamish Cleanup: Local Officials Hoping to Avoid Superfund List by Developing Own Plan," *Seattle Times,* April 20, 2000.

5 Hal Bernton, "Superfund Label Likely for River," *Seattle Times,* November 12, 2000.

6 Bernton, "Superfund Label Likely for River."

7 James P. Meador, Gina M. Ylitalo, Frank C. Sommers, and Daryle T. Boyd, "Bioaccumulation of Polychlorinated Biphenyls in Juvenile Chinook Salmon (*Oncorhynchus tshawytscha*) Outmigrating through a Contaminated Urban Estuary: Dynamics and Application," *Ecotoxicology* 19, no. 1 (January 14, 2009): 141–52, www.ncbi.nlm.nih.gov/pubmed/19685184.

8 Meador et al., "Bioaccumulation of Polychlorinated Biphenyls."

9 James Meador, interview with author, January 8, 2019; Winward Environmental, *Lower Duwamish Waterway Remedial Investigation: Task 5; Identification of Candidate Sites for Early Action* (report submitted to United States Environmental Protection Agency and Washington State Department of Ecology), Seattle, June 12, 2003.

10 James P. Meador, "Do Chemically Contaminated River Estuaries in Puget Sound (Washington, USA) Affect the Survival Rate of Hatchery-reared Chinook Salmon?" *Canadian Journal of Fisheries and Aquatic Sciences* 71, no. 1 (2014): 162–80, www.nrcresearchpress.com/doi/10.1139/cjfas-2013-0130#.XLYzwxNKjok.

11 Meador interview.

12 Mega-sites are especially large and complex cleanup areas that generally cost more than $50 million to clean up, in contrast to the majority of sites, which are confined to a small area and caused by a single outfall or source. As of 2005, fewer than 10 percent of the nation's Superfund sites were considered mega-sites. Elizabeth Southerland, "EPA Megasites," presentation at Superfund Basic Research Program Annual Meeting, San Diego, CA, January 12, 2006.

13 King County Department of Natural Resources and Parks, Anchor Environmental, and EcoChem, *Elliott Bay/Duwamish Restoration Program, Duwamish/Diagonal*

Cleanup Study Report (Final), NOAA Damage Assessment and Restoration Center Northwest, Seattle, October 2005.

14 King County Department of Natural Resources and Parks, *Duwamish/Diagonal Sediment Remediation, Dredging and Capping Operations: Sediment Monitoring, Sampling and Analysis Plan,* Seattle, October 28, 2003.

15 EcoChem and Anchor Environmental, *Duwamish/Diagonal CSO/SD Sediment Remediation Project: Closure Report*, Elliott Bay/Duwamish Restoration Panel, Seattle, July 2005.

16 Lisa Stiffler, "Duwamish Cleanup Spreads Pollutants: Large Volume of PCBs Was Taken from River, but Some Scattered," *Seattle Post-Intelligencer*, June 10, 2004.

17 Tom Paulson, "Tainted Sludge Won't Go to Tacoma: Duwamish River Sediment Will Be Sent to Klickitat County," *Seattle Post-Intelligencer*, September 11, 2003.

18 Lisa Stiffler, "Short-Term Fix for Duwamish Hot Spot: State Orders Action after Cleanup Project Exposed Pollutants," *Seattle Post-Intelligencer*, November 22, 2004.

19 Dan Cargill, interview with author, June 9, 2017.

20 SAIC, *Property Review: Terminal 117/Former Malarkey Asphalt Company*, Washington State Department of Ecology, Seattle, June 18, 2004.

21 United States Environmental Protection Agency and Washington State Department of Ecology, *Lower Duwamish Waterway Site: Seattle, Washington*, June 2003, http://duwamishcleanup.org/wp-content/uploads/2012/06/Duwamish_fs_20030627 .pdf.

22 Winward Environmental, *Lower Duwamish Superfund Site: Terminal 117 Early Action Area; Terminal 117 Engineering Evaluation/Cost Analysis (Draft)*, United States Environmental Protection Agency, Region 10, Seattle, March 4, 2005.

23 BJ Cummings, Duwamish River Cleanup Coalition/Technical Advisory Group (TAG), to Ravi Sanga, United States Environmental Protection Agency, April 7, 2005, http://duwamishcleanup.org/wp-content/uploads/2012/06/DRCCT117 EECAcomments.pdf.

24 Washington Administrative Code, title 173, chapter 340, section 170: Unrestricted Land Use Soil Cleanup Standards, Washington State Legislature, https://app.leg .wa.gov/WAC/default.aspx?cite=173–340–740; Washington Administrative Code, title 173, chapter 340, section 175: Soil Cleanup Standards for Industrial Properties, Washington State Legislature, https://app.leg.wa.gov/WAC/default.aspx?cite=173 –340–745.

25 Winward Environmental, *T-117 Upland Area Soil Investigation: Field Sampling and Data Report*, Port of Seattle, WA, July 7, 2006.

26 Robert McClure, "Tests Find High Pollution at Old Plant," *Seattle Post-Intelligencer*, March 20, 2006.

27 Craig Welch, "High levels of PCBs Uncovered near River," *Seattle Times*, November 9, 2004.

28 Washington State Department of Health, *Health Consultation: Malarkey Asphalt, Seattle, King County, Washington,* May 2001, www.doh.wa.gov/Portals/1/Docu ments/Pubs/334–239.pdf.

29 Retec, *T-117 Upland: Draft Removal Action Plan*, Port of Seattle, May 15, 2006. Residents of homes where PCBs were detected inside were advised to vacuum with a fine HEPA filter, but no other cleanup action was taken.

30 BJ Cummings, Duwamish River Cleanup Coalition, to Ravi Sanga, United States Environmental Protection Agency, April 25, 2006, in author's possession.

31 Richard Colin (chair), Sally Clark, David Della, Jan Drago, Jean Godden, Nick Licata, Richard McIvar, Tom Rasmussen, and Peter Steinbrueck, Seattle City Council, to Ravi Sanga, United States Environmental Protection Agency, May 25, 2006, in author's possession.

32 Seattle City Council to Sanga, May 25, 2006.

33 Neil Modie, "Port Opts to Exceed EPA's South Park Cleanup Plan," *Seattle Post-Intelligencer*, June 28, 2006.

34 Modie, "Port Opts to Exceed EPA's South Park Cleanup Plan." At most Superfund sites, EPA oversees cleanup studies and plans but does not develop them; the plans are crafted by the responsible parties themselves and submitted to EPA for review and approval.

35 Dan Cargill, personal communication, February 12, 2019.

36 United States Environmental Protection Agency, Region 10, *Lower Duwamish Waterway Superfund Site—Terminal 117 Early Action Area: Revised Engineering Evaluation/Cost Analysis*, Terminal 117 Cleanup, June 3, 2010, www.t117.com/documents/eeca/Final%20Revised%20EECA.pdf; Ralph Graves, managing director, to Tay Yoshitani, chief executive officer, Port of Seattle, "Declaration of Emergency," August 27, 2013, www.t117.com/documents/T117_Declaration_of_Emergency.pdf.

37 Integral, *Lower Duwamish Waterway: Slip 4 Early Action Area; Removal Action Completion Report*, United States Environmental Protection Agency, July 26, 2012, https://semspub.epa.gov/work/10/100101255.pdf, accessed April 17, 2019.

38 Integral, *Lower Duwamish Waterway: Slip 4 Early Action Area.*

39 "Poisoned Waters," *Frontline,* produced by Hedrick Smith, April 21, 2009, www.pbs.org/wgbh/frontline/film/poisonedwaters; "Boeing Company, EPA Sign PCB Cleanup Agreement," Tox-Ick.org, October 9, 2010, http://tox-ick.org/2010/10/boeing-company-epa-sign-pcb-cleanup-agreement.

40 Winward Environmental, *Lower Duwamish Waterway Remedial Investigation: Task 5.*

41 Wally Barron, interview with author, July 29, 2018.

42 "Resource Conservation and Recovery Act (RCRA) Overview," Environmental Protection Agency, February 6, 2019, www.epa.gov/rcra/resource-conservation-and-recovery-act-rcra-overview; Shawn Blocker, interview with author, February 15, 2019. EPA's manager during the plant cleanup said that decision might have been the biggest mistake Boeing ever made, describing RCRA cleanups as generally more restrictive and costly than cleanups under the Superfund law.

43 "Poisoned Waters."

44 Blocker interview.

45 Blocker interview.

46 Blocker interview.

47 Cari Simson, Maggie Milcarek, and Dan Klempner, *Duwamish Valley Vision Map and Report*, ed. BJ Cummings (Seattle, WA: Duwamish River Cleanup Coalition/ TAG, 2009).

48 Simson, Milcarek, and Klempner, *Duwamish Valley Vision Map and Report.*

49 Simson, Milcarek, and Klempner, *Duwamish Valley Vision Map and Report*, 39.

50 Duwamish River Cleanup Coalition/TAG, *Duwamish Valley Healthy Communities Project: Fact Sheet 2*, March 2013, Duwamish River Cleanup Coalition, http://duwa mishcleanup.org/wp-content/uploads/2013/03/FactSheet2.pdf.

51 Bellamy Pailthorp, "Report: More Illness, Shorter Lifespans in Duwamish River Valley," KPLU radio (now KNKX), March 27, 2013, www.knkx.org/post/report -more-illness-shorter-lifespans-duwamish-river-valley.

52 Linn Gould and BJ Cummings, *Duwamish Valley Cumulative Health Impacts Analysis: Seattle, Washington* (Seattle: University of Washington School of Public Health, Just Health Action, and Duwamish River Cleanup Coalition/TAG, March 2013).

53 Rose Egge, "Study: Duwamish Residents Have Short Life Expectancy," *KOMO News*, March 27, 2013, https://komonews.com/news/local/study-duwamish-resi dents-have-short-life-expectancy.

54 Olivia Henry, "With Focus on Toxics, Duwamish Cleanup Could Leave Other Health Problems Unsolved," *InvestigateWest*, May 15, 2013, www.invw.org/2013/05/15 /duwamish-health-impact-as-1355.

55 King County Department of Natural Resources and Parks, Wastewater Treatment Division, *Lower Duwamish Waterway Cleanup Plan Equity Impact Review*, August 2013, www.kingcounty.gov/services/environment/wastewater/duwamish-water way/~/media/FA6EDFF1E8D44AC59136D64278797333.ashx?la=en.

56 Public Health Seattle-King County comments on King County Department of Natu ral Resources and Parks Equity Impact Review, July 10, 2013, in author's possession.

57 John Ryan, testimony on Lower Duwamish Waterway Site, May 23, 2013 (in Spanish, with translation), United States Environmental Protection Agency Region 10 Public Records, Seattle; Lower Duwamish Waterway Group, *Comparison of Duwamish Clean-Up Alternatives*, original version (Seattle, 2013), in author's possession.

58 Robert McClure, "How Government and Boeing Fought to Curtail Duwamish River Cleanup," InvestigateWest, November 19, 2014, www.invw.org/2014/11/19 /the-last-days-of-the-old-1484.

59 William Daniell, Linn Gould, BJ Cummings, Jonathan Childers, and Amber Len hart, *Health Impact Assessment: Proposed Cleanup Plan for the Lower Duwamish Waterway Superfund Site*, final report (Seattle: University of Washington, Just Health Action, and Duwamish River Cleanup Coalition/TAG, September 2013), 25–29; Amber Lenhart, interview with author, February 19, 2019.

60 Linn Gould, BJ Cummings, William Daniell, Amber Lenhart, and Jonathan Childers, *Health Impact Assessment: Proposed Cleanup Plan for the Lower Duwamish Waterway Superfund Site—Technical Report: Effects of the Proposed Cleanup Plan*

on Tribes (Seattle: University of Washington, Just Health Action, and Duwamish River Cleanup Coalition/TAG, September 2013); Daniell et al., *Health Impact Assessment,* 21–24.

61 Daniell et al., *Health Impact Assessment,* 12–24.

62 Daniell et al., *Health Impact Assessment,* 34–37.

63 Daniell et al., *Health Impact Assessment,* 34–37; EcoNorthwest, *Lower Duwamish Economic Analysis,* King County Department of Natural Resources and Parks, Wastewater Treatment Division, Seattle, March 2010.

64 Jonathan Hall, interview with author, October 29, 2018.

65 Hall interview; EcoNorthwest, *Lower Duwamish Economic Analysis.*

66 Daniell et al., *Health Impact Assessment.*

67 Exec. Order No. 12898, 3 C.F.R. (1994).

68 United States Environmental Protection Agency, Office of Environmental Justice, *Plan EJ 2014* (September 2011). United States Environmental Protection Agency, Region 10, *Environmental Justice Analysis for the Lower Duwamish Waterway Cleanup: Appendix B; Proposed Plan for the Lower Duwamish Waterway Superfund Site,* Seattle, February 2013; Alexandra Gilliland, "A Review of EPA's First Environmental Justice Analysis in Conjunction with a CERCLA Remediation Plan," March 2014, Foster Pepper, PLLC, Attorneys at Law, www.foster.com/documents /a-review-of-epa-first-environmental-justice-analys.pdf.

69 United States Environmental Protection Agency, Region 10, *Environmental Justice Analysis;* Washington State Department of Health, *Lower Duwamish Waterway Superfund Site, Fact Sheet,* October 2008, www.doh.wa.gov/Portals/1/Documents /Pubs/334–139.pdf.

70 United States Environmental Protection Agency, Region 10, *Environmental Justice Analysis.*

71 United States Environmental Protection Agency, Region 10, *Environmental Justice Analysis.*

72 Gilliland, "A Review of EPA's First Environmental Justice Analysis."

73 United States Environmental Protection Agency, Region 10, *Environmental Justice Analysis;* James Rasmussen, closing remarks, 2016 Green-Duwamish Watershed Symposium, February 29, 2016, https://vimeo.com/159571959.

74 United States Environmental Protection Agency, Region 10, *Environmental Justice Analysis.*

75 Clifford Villa, personal communication, February 28, 2019.

76 The approved cleanup plan for a Superfund site is called the record of decision. Once it has been issued, EPA typically enters into legal agreements with the responsible parties, who then carry out the required studies and cleanup activities.

77 Ken Workman, personal communication, February 19, 2019.

78 Ken Workman, testimony at EPA public meeting on proposed cleanup plan for Duwamish River Superfund site, April 30, 2013, United States Environmental Protection Agency Region 10 Public Records, Seattle.

79 "The Duwamish Is My River," River for All, www.riverforall.org, accessed March 21, 2019.

80 United States Environmental Protection Agency, Region 10, *Record of Decision: Lower Duwamish Waterway Superfund Site, Seattle, WA, November 2014, Part 3 Responsiveness Summary*, 4, https://semspub.epa.gov/work/10/715975.pdf.

81 Matthew Liebman, *OSV Bold Survey Report: Puget Sound Sediment PCB and Dioxin 2008 Survey*, United States Environmental Protection Agency, September 11, 2008.

82 Simson, Milcarek, and Klempner, *Duwamish Valley Vision Map and Report*.

83 AECOM, *Executive Summary: Final Feasibility Study; Lower Duwamish Waterway, Seattle, Washington*, Lower Duwamish Waterway Group, October 31, 2012, 28, 38.

84 AECOM, *Executive Summary: Final Feasibility Study*, 28, 38; James Rasmussen, BJ Cummings, and Lee Dorrigan, Duwamish River Cleanup Coalition/TAG, to Allison Hiltner, United States Environmental Protection Agency, January 14, 2011, EPA Region 10 Public Records, Seattle, WA.

85 Dan Cargill, personal communication, February 12, 2019.

86 Lideos, *Green-Duwamish River Watershed: Compendium of Existing Environmental Information*, Washington State Department of Ecology, Toxics Cleanup Program, October 2014, 33–35.

87 Lideos, *Green-Duwamish River Watershed*, 35–40; Dan Cargill, personal communication, February 12, 2019.

88 King County representative, Health Impact Assessment Liaison Committee meeting, McInstry Innovation Center, March 11, 2013.

89 "Poisoned Waters."

90 United States Environmental Protection Agency, Region 10, *Record of Decision: Lower Duwamish Waterway Superfund Site*, Seattle, November 2014, https://semspub.epa.gov/work/10/715975.pdf; Northwest Regional Office and Lideos, *Lower Duwamish Waterway Source Control Strategy*, Washington Department of Ecology, Toxics Cleanup Program, Pub. No. 16-09-339, June 2016.

91 Sarah Kavage and Nicole Kistler, interview with author, March 20, 2019.

92 Jim Demetre, "A River Revived: Forty Artists from around the World Celebrate the History and Future of the Duwamish," *CityArts*, June 23, 2015, www.cityarts magazine.com/river-revived.

93 Demetre, "A River Revived"; Duwamish Revealed opening ceremony, video recording, June 5, 2015, https://vimeo.com/161254322.

94 Demetre, "A River Revived."

95 Sophorn Sim, interview with author, March 29, 2019; Russell Ross, *Cambodia: A Country Study* (Washington, DC: Federal Research Division, Library of Congress, 1990), www.loc.gov/item/89600150.

96 Sam Le, "Celebrating the Cultural Importance of Water," *Northwest Asian Weekly*, August 3, 2018, http://nwasianweekly.com/2018/08/celebrating-the-cultural-impor tance-of-water.

97 Sim interview.

98 Sim interview.

99 Isa Kaufman-Geballe, "Translating Sustainability," *Planet Magazine*, December 12, 2017, https://theplanetmagazine.net/translating-sustainability-49302a9bf3b3.

100 Sophorn Sim, address at "Rivers: An International Storytelling Event," Sullivan Community Center, Tukwila, WA, March 31, 2019.

101 Sally Macdonald, "Belulah Maple Norman, 98, Artist, Granddaughter of Pioneer Family," *Seattle Times*, January 11, 1992; Louise Ann Jones, interview with author, May 24, 2018.

102 Patrick McRoberts, "Hugo, Richard (1923–1982)," HistoryLink, January 20, 2003, www.historylink.org/File/5082; Richard Hugo, *A Run of Jacks* (Minneapolis: University of Minnesota Press, 1961).

103 Gene Gentry McMahon, interview with author, March 29, 2019; River for All, www.riverforall.org, accessed March 26, 2019; Duwamish River Artist Residency, www.duwamishresidency.com, accessed March 25, 2009.

104 Clair Gebbin, "Ancient Traditions Shared in Duwamish Cultural Longhouse," *Mercer Island Reporter*, October 1, 2009, www.mi-reporter.com/news/ancient-traditions-shared-in-duwamish-cultural-longhouse; Steve Griggs, *Listen to Seattle*, http://listentoseattle.blogspot.com, accessed March 26, 2019.

105 Paulina Lopez, interview with author, March 15, 2019.

106 Lopez interview.

107 Carmen Martinez, interview with author, May 4, 2018.

108 Roger Fernandes, interview with author, March 30, 2019; Daniella Cortez, interview with author, March 30, 2019.

109 Joel Connelly, "A New South Park Bridge: The Neighborhood Made It Happen," *Seattle Post-Intelligencer*, June 29, 2019; Ken Workman, personal communication, March 30, 2019.

110 Fernandes interview.

111 Ryan Calkins, address given at Duwamish Valley Youth Corps Mural Unveiling, Duwamish Waterway Park, Seattle, WA, March 30, 2019.

112 Roger Fernandes, address given at Duwamish Valley Youth Corps Mural Unveiling, Duwamish Waterway Park, Seattle, March 30, 2019.

113 "Lower Duwamish Waterway," Superfund Site, United States Environmental Protection Agency, October 20, 2017, https://cumulis.epa.gov/supercpad/SiteProfiles/index.cfm?fuseaction=second.Cleanup&id=1002020; Alison Morrow, "Seattle Recycling Company Settles Lawsuit over Duwamish River Pollution," *King5 News*, January 18, 2019, www.king5.com/article/tech/science/environment/seattle-recycling-company-settles-lawsuit-over-duwamish-river-pollution/281-424a3e90-8982-40c8-8c13-070ab42a7a16; United States Environmental Protection Agency, Region 10, *Fishing in the Duwamish River*, Seattle, WA, February 2017, https://semspub.epa.gov/work/10/100046881.pdf.

114 City of Seattle, Duwamish Valley Program, *Duwamish Valley Action Plan: Advancing Environmental Justice and Equitable Development in Seattle*, Seattle, June 2018.

115 Gilliland, "A Review of EPA's First Environmental Justice Analysis"; James Rasmussen, interviews with author, 2015–19.

116 Rasmussen, closing remarks at 2016 Green-Duwamish Watershed Symposium.

117 United States Environmental Protection Agency, Region 10, *Environmental Justice Analysis;* "About the Roundtable," Lower Duwamish Waterway Roundtable, n.d., www.duwamishwaterwayroundtable.org/about-the-roundtable, accessed April 18, 2019.

118 Jonathan Hall to Julie Congdon, United States Environmental Protection Agency, and Sophie Glass, Triangle Associates, January 10, 2018; Jonathan Hall, personal communication, February 28, 2019.

119 Alberto Rodriguez, personal communication, February 28, 2019.

120 James Rasmussen, personal communication, February 28, 2019.

A NOTE ABOUT SOURCES

A great deal of the material contained in *The River That Made Seattle* comes from published histories of Washington State, Puget Sound, King County, and Seattle, reframed here with a specific focus on the Duwamish watershed and viewed from the perspective of its Native and immigrant communities.

Until now, only two books about the Duwamish River have been published—*The Price of Taming a River*, by Mike Sato (1997) and *The Once and Future River*, a collection of photographs by Tom Reese and essays by Eric Wagner (2016). In addition, *The Duwamish Diary* was written in 1949 by a group of Cleveland High School students and their teacher, casting the river itself in the role of narrator. These have been foundational texts for my research into the history of the Duwamish River and the people who shaped its course, along with three contemporary environmental and Native histories of the Seattle area—*Emerald City*, by Matthew Klingle; *Native Seattle*, by Coll Thrush; and *Chief Seattle*, by David Buerge—along with accounts by Seattle's early pioneers and their early-twentieth-century biographers and chroniclers, including David and Eliza Denny, Clarence Bagley, Thomas Prosch, and Robert Nesbit.

I have also relied heavily on the excellent historical essays on the HistoryLink website and on newspaper accounts preserved at the University of Washington Libraries, museums, government archives throughout King County, and the Duwamish Tribe's historical archives at their Longhouse and Cultural Center in Seattle. I consulted lesser-known books produced by historical associations and academic institutions; journal articles; organizational and topical websites; government documents, particularly those of the Bureau of Indian Affairs, the Environmental Protection Agency, the Washington State Department of Ecology, King County, the Port of Seattle, and the City of Seattle; and federal and state census, birth, death, and probate records obtained through Ancestry.com and other sources. I am especially indebted to my research assistant, Jennifer Smith, a PhD candidate in history at the University of Washington, for her help in locating and organizing much of the archival material used in my research.

This book would not have been possible without the generosity of many people with personal and professional connections to the Duwamish River, who have provided me with family documents that were not available in the historical record. At the top of this list is James Rasmussen, Duwamish Tribe member, third-generation Duwamish Tribal Council representative, and founding director of the Duwamish Tribe's Longhouse and Cultural Center. James and his sister, Virginia Nelson, have provided me with access to extensive family photographs and correspondence, as well as to genealogical and property records that date to the mid-1800s. These records trace the family's tradition of tribal leadership predating European settlement of the Puget Sound region and often contradict other historical accounts, written by white settlers, about Indigenous life and families during the early years of colonization.

Duwamish tribal and family records were also provided by Ken Workman, a former Duwamish Tribal Council representative and past president of Duwamish Tribal Services, and by Duwamish Tribe chairwoman Cecile Hansen and her family. Each of these individuals provided written transcripts of their families' oral histories and were kind enough to allow me to interview them. I conducted additional interviews with numerous Duwamish Valley immigrant residents and their descendants, government officials, industry representatives, artists, and others with a connection to the story of the Duwamish River.

Finally, I have worked as an environmental consultant and advocate on the Duwamish River for twenty-five years, and I explored much of the current and historic watershed myself while writing this manuscript. Much of the material in the book comes from my own experience and accumulated knowledge of events along the river from 1994 to 2019. The manuscript includes parts of my own story where it illuminates or provides necessary context for events.

INDEX

Blocker, Shawn, 143–44

Boeing, William, 85–86

Boeing Airplane Company: airfield used by, 116, 151; description of, 86, 88–89, 95; Duwamish historic village and, 116, 151; Environmental Coalition of South Seattle and, 160; industrial waste from, 102, 104, 129–30; Maple's employment at, 163; PCBs at, 142, 144; Plant 2, 141–45; pollution by, 142–44; Workman's employment at, 155

Boeing Field, 82, 87, 126, 142

Boldt, George H., 109–11, 112

Bolton, Andrew, 45

bone game, 55–56

Bowman, Elsa, 77–78

Boyle, Bob, 114

Bray, Mike, 100–101

breweries, 82–83, 171

brick factories, 83

Brno, Tanya, 160

Buerge, David, 27, 44

Burke, Thomas, 61, 63–64, 68

C-2 float planes, 86

Calkins, Ryan, 168

Camas, 20, 40

Cambodia, 161–62

Camp Minidoka, 94

Campbell, Sarah, 6

canal: Beacon Hill portion of, 64–65, 68; Burke's involvement in, 61, 63–64; Congressional testimony regarding, 66; dredging of, 78–81; Duwamish Waterway Commission, 76–81; extension of, 89; government of, 66; Hamm's involvement in, 77; Lake Washington Improvement Company, 59–60; lawsuits regarding, 65, 79–80; Mercer's involvement in, 61; North, 63–64, 66–67; photograph of, 59; Pike's involvement in, 61–62; progress of, 79–81; public funding for, 67; Ross's

involvement in, 59–60, 68; Semple's involvement in, 62–66; tide-flat, 66, 70–71. *See also* Ship Canal

Cargill, Dan, 119, 134, 140, 158

Carleton Street, 85

Carlson, Lucille, 89–90

Cascade Mountains, 30

Cavanaugh, Martin, 30

Cavanaugh, Mary Ann, 29–31

Cedar River: dam on, 90; description of, 10–11; Duwamish people living near, 50; Jack Bigelow and, 54; rerouting of, 74; sawmill on, 38; settlements on, 39; settlers on, 23, 30

Centennial Flour Mill, 65, 75, 88

chemical pollution, 105

ChemPro, 103–4

Chi, Ruben, 161–62

Chicago World's Fair, 87

Chief Seattle. *See* Se'alth

Chief Seattle and the Town That Took His Name, 27

Childers, Jonathan, 148

Chinook Wind, 17

Chiyoda Corporation, 114

City Planning Commission, 89

Clark, Marianne, 84–85

Claussen-Sweeney, 82

Clean Water Act, 9, 106, 114, 116

Cleveland, Grover, 63, 76

Clinton, Bill, 152

coal mining, 37–38, 44, 56, 100, 102, 103

Collins, Luther: description of, 11, 28–30, 33, 35, 37; land ownership by, 39–40; lynchings by, 40

Columbia River, 37, 90

Columbian, 39

Colville River, 44

combined sewer outflow, 132–34, 142, 157

Commencement Bay, 16, 78

commercial fishing, 90–91

Coney Island, 80

Congdon, Julie, 170

Peck, Norm, 101–2
Peterson, Piper, 141
Pew Charitable Trusts, 148
Pierce County, 70, 72
Pike, Harvey, 61
Pike, John, 61
Pike Place Market, 87, 99
Pioneer Square, 76
Pitkin, Stanley, 108–9
Place of the Fish Spear, 40
Point Elliott, 41, 43, 50
Point No Point, 41–43
Point Rediscovery, 120–21
pollution, 12; air, 96–98, 148; asphalt, 135–37; by Boeing Airplane Company, 142–44; chemical, 105; in Duwamish River, 12, 115–16, 121, 124–25, 130–32, 134–41, 144; garbage burning, 96–98, 100; hazardous waste, 101–3, 105; health effects of, 147; Hudson River, 115; industrial waste, 102–4, 112, 136; Lower Duwamish Waterway, 148; Lower Duwamish Waterway Group recommendations, 157–58; PCBs, 113–14, 122, 127–29, 132, 136–40; salmon affected by, 122–24, 127; sewage, 105, 107–8, 112; TBT, 122–24; Terminal 117, 135, 137, 139–42; toxic, 101, 104, 114, 122, 125, 130–31, 146; upriver, 157–59; water, 105–6
polychlorinated biphenyls. See PCBs
Porter, A. L., 45–46
Port Madison, 21, 32, 47–48, 50, 52
Port Madison Indian Reservation, 32, 50
Port of Seattle, 3–5, 6, 113, 137–38, 140–41, 169
Premel, Albert, 84
Proctor, Gerald, 54
Prohibition, 77
Puget, Peter, 19
Puget Mill Company, 80
Puget Sound, 8, 17; chemicals in, 124; estuaries in, 127, 129; fish stocks in, 90; fishing in, 106; Hudson Bay Company trading post in, 27; immigrants in, 27; Lake Washington and, canal efforts between, 62; reservations in, 25, 42; salmon in, 90, 92, 109, 113; settlers in, 26–27, 43, 45, 53; social organization in, 21; trading in, 21, 39; tribal confederacy in, 21; tribes of, 20–21, 23, 46
Puget Sound Fire Clay Company, 83
Puget Soundkeeper, 115–16, 121, 135. See also Puget Sound(keeper) Alliance
Puget Sound(keeper) Alliance, 8–9, 132, 168–69. See also Puget Soundkeeper
Puyallup River: description of, 48, 70; fishing on, 108–9
Puyallup Tribe, 48, 56

Quartermaster Harbor, 52
Queen City Landfill, 102
Quio-litza, 25–26, 38, 52, 55–56, 76, 93, 119

Rainier Brewing Company, 82
Rasmussen, Ann, 25–26, 57, 120, 174
Rasmussen, James, 24, 38, 52, 119–20, 154, 161, 164, 169–72
Ravenna, 73
Ray, Dixie Lee, 6
refugees, 161
Renton, 38, 83
Renton Clay Company, 83
reservations: Native people on, 49; in Puget Sound, 25, 42
Resource Conservation and Recovery Act, 143
Restoration Point, 19
Revelle, Randy, 113
rezoning, 98–100
River for All campaign, 156, 164, 167
Riverkeepers, 115
Rivers and Harbors Act, 4, 62, 105, 114–15
Roaring Rock Creek, 10
Robert Wood Johnson Foundation, 148

tribe(s): EPA cleanup plan effects on, 150; fishing rights of, 107–10, 150; health as defined by, 150. *See also specific tribe*
Truman, Harry S., 105
Trump, Donald, 175
Tsu-Cub, 21–22
tsunamis, 17, 94
Tukwila, 93
Tukwila Heritage and Cultural Center, 163
Tulalip Reservation, 50
Tupt-Aleut, 18, 23–25, 38, 52, 56, 120
Tuttle, Abner, 52–53
Tuttle, Nellie, 52–53, 56–57, 67, 120
Tuwaltwx, 35
Twins Patrol the River, The, 31

Uhlman, Wes, 6
Ungaro, Louie, 111
Union City, 61
United States Court of Appeals, 110
University of Washington: description of, 6, 57; School of Public Health, 148
upriver pollution, 157–59
US Environmental Protection Agency, 101
US Postal Service, 86

Valley Daily News, 103
van Asselt, Henry, 10, 30, 35, 83
Vancouver, George, 18–21
Vancouver Island, 18, 24, 36
Vann, Song, 161–62
Varanasi, Usha, 122, 124, 127
Vashon Island, 52–53
Villa, Clifford, 155
Villard, Henry, 65

Wagener, O. F., 71
Wagner, Paul Cheoketen, 160

wapatoes, 37
Washington State Sportsmen's Council, 91, 107
Washington State Supreme Court, 108
Waste Action Project, 116, 121, 137, 151
Water Festival, 161
Waterman, T. T., 23, 41
Waterman, W. H., 50–51
water pollution, 105–6
Water Pollution Control Act, 106, 114
West, Ron, 103
Western Processing, 102
Westneat, Danny, 103–4
West Waterway, 111, 123
Whidbey Island, 18, 23, 25, 34–35
White River: color of, 14; description of, 10–11, 48; Duwamish people living near, 23, 43; flooding at, 71–72; Naches Pass, 45; rerouting of, 72, 75; Se'alth's battles near, 22; settlers on, 30; treaties and, 43
White River Commission, 72
Whulge, 17, 19, 21, 33, 43. *See also* Puget Sound
Williams, David, 62
Wilma, David, 109
Wingard, Greg, 100–101, 116, 121, 137–38, 151
Wong Tsu, 86
Woodinville, 10
Woodward, Bill, 99
Workman, Ken, 155–56, 167, 174
Wyamook, 36

Yakama Tribe, 25, 44, 47
Yamaguchi, Catherine, 94
Yamaguchi, David, 94
Yamaguchi, Tadashi, 93–95
Yesler, Henry, 30, 37, 46, 58
Yuliqwad, 7

ABOUT THE AUTHOR

BJ Cummings founded the Duwamish River Cleanup Coalition in 2001 and is the community engagement manager for the University of Washington's Superfund Research Program in the Department of Environmental and Occupational Health. She has coauthored several community health studies, including the *Duwamish Valley Cumulative Health Impacts Analysis* and *Duwamish River Superfund Cleanup Plan Health Impact Assessment*. Cummings served as the Puget Soundkeeper Alliance's "Soundkeeper" from 1994 to 1999 and as Sustainable Seattle's executive director from 2016 to 2018.

Cummings holds a master's degree in environmental geography from UCLA and is the author and producer of numerous articles, books, and documentary films on environment and development issues in the Seattle area and throughout the Americas. Over the past two decades, Cummings has been recognized as a National River Network River Hero, a Sustainable Seattle Sustainability Hero, a King County Green Globe winner for environmental activism, a recipient of Puget Soundkeeper Alliance's Inspiration Award, and one of *Seattle* magazine's ten most influential leaders.